WRITING
TECHNICAL ARTICLES,
SPEECHES,
AND MANUALS

WRITING TECHNICAL ARTICLES, SPEECHES, AND MANUALS

MARK FORBES

WILEY

John Wiley & Sons, Inc.

New York • Chichester • Brisbane • Toronto • Singapore

Publisher: Stephen Kippur
Editor: Therese A. Zak
Managing Editor: Andrew Hoffer
Designed, edited, and composed by
Publications Development Company

Library of Congress Cataloging-in-Publication Data

Forbes, Mark
    Writing technical articles, speeches, and manuals / Mark Forbes.
        p.   cm.
    Bibliography: p.
    Includes index.
    ISBN 0-471-60096-2 (pbk.)
    1. Technical writing.   I. Title.
    T11.F58   1988
    808′ .0666021—dc19   88-10157
                                                                    CIP

Printed in the United States of America

88 89 10 9 8 7 6 5 4 3 2 1

PREFACE

If you are like most technical people, you bought this book because writing is not extremely enjoyable; maybe it is even difficult for you. Well, you are not alone. I have never disliked writing, but I did not begin to really enjoy it until I developed a systematic approach to writing . . . sort of a scientific method. Using this method, very large writing problems, as with very large engineering problems, reduce to a series of manageable tasks. The final task, of course, is to integrate the subparts into a completed work.

The biggest obstacle to a writing project is actually getting started. Again, reducing the problem to subparts is the place to begin. From there, the solution simply builds on itself. The easiest way to get started is with an outline. An outline acts as your preliminary specification. I will show you how to effectively use an outline for whatever problem you are trying to solve.

There are basically three types of technical writing: Articles, including data sheets, application notes, and magazine articles; speeches, including "forum"-type speeches and technical presentations for small groups; and technical manuals of all types. The technical manual is generally the most difficult, because the audience is generally at a much lower level of technical competence than in the other two classes.

Each of the three areas is covered in its own section. Some techniques apply across the board, and others are specific. By applying these techniques, you should find your writing tasks greatly simplified.

Finally, you will find this to be a realistic book, for real-world people. I could easily suggest that you start that speech six weeks before it is due, and how to do research on it in that amount of time. But face it, you are not likely to follow that suggestion. We all sometimes wait until the last minute because there are always seemingly higher priority tasks to do first. Being realistic, I have tried to offer pointers for what to do when you are out of time, in addition to the "proper" way of preparing material for presentation.

That is what you will find in this book, in a nutshell. Treat writing as any other technical problem, and you will find it easier, quicker, and more enjoyable.

ACKNOWLEDGMENTS

Listing the author's name on the spine of a book is like listing the quarterback of a football team. Writing and producing a book is a team effort. I would like to take this space to offer my deep appreciation and thanks to the key members of the team that helped me get this book together.

Tom Collins helped me write Chapters 3 and 4, and also helped edit all of the first four chapters. Perhaps even more important was the fire that he lit under me when my motivation was running low. As my senior editor at *Computer Technology Review,* he also taught me a great deal about editing. Without Tom, this book may never have happened.

Teri Zak is the Editor at John Wiley who liked my idea and stayed with me even when the going got tough. I'm sure Teri invested some extra money on aspirins the times I was late for a deadline. But more importantly, she believed in me even when I was doubting.

Cyndy Drummey, on more than one occasion, dropped whatever she was doing to help type or research a fact for me. I wish I had more friends like her.

And finally, I want to thank my wife, Glenell, and my two children, Kirk and Lindsay. They had to put up with me being stuck in front of my word processor for many weekends, and often late into the night. Now, we can go sledding!

TRADEMARKS

Multibus is a trademark of Intel Corporation.

WordStar is a trademark of MicroPro International Corporation.

RightWriter is a trademark of RightSoft Corporation.

Heathkit is a trademark of the Heath Corporation.

Microsoft is a trademark of Microsoft Corporation.

CONTENTS

Contents

S E C T I O N
ONE

**ORGANIZING
YOUR
IDEAS**

CHAPTER

1

THE OUTLINE:
A SPECIFICIATION
FOR WRITING

"Writing is a pain in the butt."

Writing is a pain in the butt." As editor of a computer trade magazine, I have heard that remark from too many people to be able to give proper credit for it. The sad thing is, though, that writing—*especially* technical writing—is difficult for most people. The good news is that writing does not have to be difficult by definition.

Writing is often made more difficult than it has to be—particularly by writers themselves—though English teachers and other professionals also share the blame for popularizing the difficulty of writing. We have all heard complaints about the technical complexity of some products. Often we, as technical people ourselves, have a hard time understanding why someone is confused about a product that seems simple enough to us. The product works well in the lab, and we never have any trouble operating it.

In some instances, familiarity with a product coincides with an intimate understanding of a complex set of physical variables that a nontechnical person may not be conversant with. In other cases—and this is true more often than not—the product is complex because the designers have communicated the directions for its operation poorly. It may be that the directions are loaded with technical jargon that has no reality for the user. Alternatively, it may be that the writers presume knowledge that the user does not have.

Many times I have spent hours struggling with an instruction manual, only to stumble onto the correct method of operation, which turned out to be trivial. The key to imparting a clear understanding of a product or procedure usually is not inherent in the material itself, but in the manner and language used to present that material.

When you are trying to explain a detailed technical subject to a nontechnical friend, you probably do so by using analogies. By explaining the subject in terms that your friend is familiar with, he or she is much more likely to understand at least the general concept involved. "Five megabytes per second" might be totally meaningless while "five million characters per second" or "a million words per second" would be understandable to nearly anyone.

If you explain an idea in terms that another person is familiar with and can understand, that idea will be more effectively communicated. In writing this book, I set out to communicate ways of making writing easier for engineers, technicians, and other technical people. My approach is to communicate in your own language . . . that is, by demonstrating that technical writing is analogous to technical design.

WRITING BY DESIGN

Imagine that you have been assigned a complex design project, such as a large computer system. The initial definition might be "Design a very fast computer that can handle 32 users." While that is a definition, it isn't nearly complete enough for you to begin the actual designing of the system. To a nontechnical person, that may seem an almost insurmountable task. An experienced electronic engineer, especially one who has designed similar systems in the past, knows that even though the task is large and complex, it can certainly be accomplished.

How do you begin a task as large as this one though? It turns out that *how* you start is extremely important. The only way to effectively handle a problem of this magnitude is to break it down into smaller problems. You begin this by defining all of the questions that need to be answered to solve it.

For example, for the basic design task just posed, there are a multitude of questions that have to be answered, such as: How fast is fast? What keyboard is preferred? How much memory is required? What are the cost constraints? Are the users all in one location? What space is available for equipment? What is the load factor (how many simultaneous users)? What output device is needed? Is more than one operating system to be used?

When you have posed all of the questions you can think of, you are ready for the next step. Next come the many meetings that answer these questions and perhaps create more by clarifying what this computer will do, who it will be designed for, and what it will cost. During these meetings, a preliminary design specification is assembled. Some of the specifications are well understood and are almost immediately apparent. Other specifications are not well known, or are not within the expertise of the engineering and marketing departments.

The engineering department builds the research foundation upon which development will take place. The marketing and sales departments research markets, competition, possible volumes, and potential customers. Finance works in conjunction with engineering and marketing to determine costs and whether it is profitable to build such a product. Manufacturing supplies data about the reproducibility of any given design.

After many design review meetings, a final specification is written. Only now can the actual design begin, since (with any luck) all the questions were answered and a clear, concise document was produced to serve almost as a set of instructions.

This document is the final specification, which was developed by spending a great deal of time and money to expand the preliminary specification. Some parts of the preliminary specification may be virtually unchanged, and others may have been changed completely during this evolution. By the time the final specification is produced, the designers have a very clear idea of what the solution is, and how to arrive at its design . . . the specification has been so well thought through that the actual design becomes relatively easy.

Adapting the Design Process to Writing

Although no one would begin designing a complex system like a computer without a great deal of research and investigation, most of the time that is exactly how people write technical papers and manuals. However, just like the product design process, the original project definition may leave many questions unanswered. Most of these questions will get answered as you write, but the logical flow may be poor; you may have left out something important, or perhaps worse, you may have included extraneous information that dilutes or obscures your true point.

The design process just described is basically an adaptation of the scientific method. The scientific method can also be adapted for writing—or just about any other task, for that matter.

The first step in beginning a writing project is to define the described task or problem, much as was done for the computer to be designed. For example, when I first set out to write this book, my definition was to write "a book that describes technical writing in engineering terms." Clearly, I had the idea, but I needed more detail before I could begin the project (not to mention that the publisher needed more convincing!).

The Outline: The First Step in Organizing Your Ideas

The next step is to write the specification. The specification for a writing project is contained in an outline. As in the design process, a specification is arrived at through an iterative process. The preliminary specification is still broad, but more refined than the original definition. Then that preliminary specification is developed into the final specification from which the design is actually executed.

For most technical people, this procedure is totally familiar. It is *the* method for designing almost any type of product. However, the same people fail to extend this logical, structured, problem-solving procedure beyond their day-to-day work.

The outline is the analogy of the product design specification. Like the specification, it will provide the framework around which your entire written work will be built. The more thorough you are at this stage, the easier the product will come together—and, generally, the better the quality of the final design. The same holds true for the relationship between preparing the outline and writing a final manuscript.

If you were like me when you were in school, you didn't particularly care for writing outlines. Most of the time, I wrote the outline *after* the paper was complete. This satisfied the teacher (since she did not know when I actually wrote it), but it sure didn't aid the writing process at all.

For this book, my preliminary outline was simply a list of the topics that I thought it was necessary to cover. My first outline for this book appears in Figure 1. This outline lacked details, and I certainly could not write the book very well from such a vague set of topics. This preliminary listing was followed by expansion of the outline where the details were filled in.

At this point in the process, with only a bare listing of key elements to cover, I could expand the outline somewhat without

> I. Organizing
> II. Writing Articles
> III. Writing and Giving Speeches
> IV. Writing Manuals

Figure 1
This was the first rough outline for the book.

much research. However, it did take thought. After deciding what topics were necessary, which were not, and how the topics should be ordered, I found the first expansion of my outline had become that shown in Figure 2. This outline is still rather condensed, but the thought processes determined what was essential and how it should be ordered. I also left out items that might be distracting, such as how to file, what outline style to use, and so on.

After the initial expansion, you will want to go on to expand the outline to as much detail as possible. The advantage of this process will be shown in a short while. There are basically two methods to use in expanding your outline. The one you choose to use depends on the topic, and how familiar you are with it.

Expanding the Outline

If you are quite familiar with the topic, you can progress through the outline expanding it all at once. This also allows you to rearrange the topics if necessary to improve the flow of information. As you are expanding the outline, you may find it necessary to research some of the topics to fully expand that topic.

If you are writing about a topic that is not too familiar to you, expanding each topic individually is probably the best choice. This allows you to concentrate your efforts on only one

```
I.  Organizing
    A.  The Importance of an Outline
    B.  Research and Collecting Data
```

Figure 2
The first expansion of topic I of the outline.

subject at a time. The importance of expanding one topic at a time is that you will need to absorb a lot of material about the subject. If you choose this method, you will want to review your outline after you have finished expanding it to determine if your original thoughts about the subject were accurate. Often, you will want to change the order, or even add or eliminate topics after you have learned more about the subject. The mechanisms and procedures for researching topics are covered in Chapter 2.

After more thought, and some research, the final expansion for Section One of this book became the outline shown in Figure 3. Not only is this significantly more detailed than any of the earlier steps, it is the outline that I actually used to compose this book.

I. Organizing
  A. Analogy to the Design Process
    1. Preliminary Specification
      a. It's a general definition
      b. Lacks detail and needs research
    2. Final Specification
      a. Evolves from preliminary spec
      b. Makes actual design easy
  B. The Importance of an Outline
    1. Preliminary Spec for Writing
    2. Expanding the Outline
      a. Expanding as a whole
      b. Expanding by topic
    3. First Step in Organizing Ideas
    4. Makes Writing Easier
    5. Keeps You from Straying from Subject
    6. Review of Outline Saves Time When Writing

Figure 3
The final outline for Chapter 1 of this book.

Writing Becomes Easier

There is a saying that a writer is working hardest when he or she appears to be doing nothing. I can say from experience that this is indeed true. As with an engineering product design or an architect's conception of a building, thinking about what you are going to write is much more difficult than actually pushing the keys on the word processor or typewriter.

Motivating you to think carefully is perhaps the most beneficial aspect of using an outline, and is again analogous to a technical design. As you develop the outline, you really have to think about the subject. At first, this thinking may take the form of questions. Then, as your research answers your questions, you begin to formulate ideas and combine them.

We have all seen the Hewlett-Packard "What If?" television commercials where the engineer pulls into a phone booth or jumps out of the shower wearing only a towel—just so he can call his boss and tell him his great idea for solving whatever problem he has been thinking about. Those commercials convey a point: Ideas come at all times and places. I seem to do some of my best thinking while in the shower. Try to find that place or atmosphere most conducive to your thinking processes.

By the time you finalize the outline, you have taken your preliminary ideas and combined them with your research to assemble a body of knowledge about the entire subject. Getting to this point has required a lot of detailed thought. The "thinking through" process continues as you reorganize the outline for a better flow of information.

Once you have determined how to "package" your work, it is time to do the actual writing. With all the research and thinking you have already done, you have essentially written the paper in your head. Getting it on to paper is now more of a transcription process than the very laborious task it would be if you jumped right in and started writing without the outline.

This is the same procedure you would use in a design. You would never go to a model maker and say "make me a gear." Thought, planning, and design have to be done first. The results of the preliminary work—"make me a 3 inch diameter, .025 thick stainless steel gear with 24 teeth" and so on—are far more specific and helpful to the model maker than the first generality.

You could write first and refine it later, but you will not be satisfied with the results. Putting in the time up front yields a more elegant piece of writing that does not take nearly as much time to finish.

I. Organizing
  A. Analogy to the Design Process
    1. Preliminary Specification
      a. It's a general definition
      b. Lacks detail and needs research
    2. Final Specification
      a. Evolves from preliminary spec
      b. Makes actual design easy
  B. The Importance of an Outline
    1. Preliminary Spec for Writing
    2. Expanding the Outline
      a. Expanding as a whole
      b. Expanding by topic
    3. First Step in Organizing Ideas
    4. Makes Writing Easier
    5. Keeps You from Straying from Subject
    6. Review of Outline Saves Time When Writing

Figure 4

Note how complete the final outline is. Just about all that is necessary is to make paragraphs from the topical list to get to the final form of the chapter.

Look once again at the final outline for the first section of this book, shown in Figure 4. Compare the outline with the actual contents of this chapter. You will notice that about all I had to do to get from the outline to the finished chapter was to make paragraphs from the topic list of the outline. All of the ideas were already in place and well thought out, making the actual writing much easier.

Stay on the Subject

Another benefit of a carefully developed outline is that it helps to keep you on the subject. You can have a fantastic idea, develop it logically, and make an astounding conclusion. If you lose the readers in details and take them down rabbit trails, you may camouflage your point so well that the reader misses it.

I used to work for a man who was very intelligent and had an innate understanding of business. He could tell fascinating stories about business and marketing. The problem was that after two hours of listening to his anecdotes, everyone in the staff meeting had forgotten that we were gathered together to increase sales in the Northeast. While I learned a lot about business and philosophy, we never did get those sales up.

Sometimes an outline will require you to make a difficult choice. There may be valid reasons for straying from your outline, and quite often topics are added after you have begun writing. When you notice that you have strayed from your plan, stop and determine if it is really necessary and not distracting to do so. I have wiped out many wonderful paragraphs that I wrote which, unfortunately, didn't support the topic or diluted the argument.

With most technical writing, you are constrained with respect to length. The problem this usually presents is that you have too little space to present the material that you had intended to. If you stray away from your point, not only do you dilute its impact, but you have stolen your most precious resource—space—from your primary objective.

Keep the Outline Handy

It is not good enough to write out the outline and forget it. You have to keep referring to it to remain on track and to cover everything, and in the order you had planned. It is as important to change or update the outline as you write. Referring to my preliminary outline, you will notice that I reversed the order of the sections on writing a manual and writing and giving speeches.

As I began doing more detailed research, I realized that while writing a speech and writing an article have much in common, writing a technical manual presents a number of unique problems. Thus I felt that it made better sense to follow the section on technical articles with the one on manuals, then leave the speech writing to the end to "wrap up and tie things together." Use the outline to make the writing easier, but don't let the outline handcuff you.

GOOD WRITING FOLLOWS GOOD DESIGN

It should be pretty clear that an outline can be a real boon when it comes to writing. If you are not convinced of this, you will be as soon as you have developed your first outline and actually followed it through to the successful completion of a manuscript. I have used this book for all of the examples in this chapter, however, an outline is useful regardless of the length of the project—a book, an article, or a manual. I even have begun to use outlines when writing personal letters. Proposals, grants, business plans, and other business-related writing can be much improved by jotting down a quick outline before beginning.

The steps in designing a good piece of writing are exactly analogous to any technical design. Begin with a broad definition. Then construct a preliminary specification in the form of an outline.

Using the outline, research and expand it until you have reached a level of detail at which you feel comfortable to begin the actual writing. An overly detailed outline is preferable to one that lacks detail—especially since the reason it lacks detail at some specific point is almost certainly because you have not thought sufficiently about what you want to say at that point. Don't be afraid to rearrange the outline to clarify the flow and the logic of your article, even during the project.

When you have reached the point of actually writing, you will have already "written" it in your mind. Transcribing it to paper will surprise you with how short of a time it takes when compared to "writing on the fly."

Once you have finalized the outline, refer to it frequently and stick with it unless you can genuinely improve it. If you wander too far afield, take the ax to it! Stay focused and refer to your outline frequently. Like good product designs, good writing results from planning, research, and logic.

CHAPTER
2

**COLLECTING
DATA**

*"The Outline not only keeps
your writing from straying, it
keeps your research from straying."*

*I*n addition to the help that an outline provides during the writing process, it also facilitates the research portion of your project. Once the outline is written, it is usually quite clear in what areas you have adequate knowledge, and in what areas you need to do some research to discuss the topic adequately.

Once you begin the research, the outline helps you to focus on just the topics you will cover. The outline not only keeps your writing from straying, it keeps your research from straying.

DETERMINE AREAS NEEDING RESEARCH

Using the outline as a guide, select the areas that you have the most difficulty in fleshing out. What would you need to know in order to discuss those areas adequately? If the outline does not indicate precisely how you will make the points necessary to explain the major heading or subheading, you probably need more thought, more information, or both.

Researching one topic at a time will be the easiest. Remember, though, you created your outline and you can expand and change it as necessary during the research process. An outline is a work tool, to be altered, adjusted, and modified to better meet your needs. How you compile your research material is a personal matter. Research papers written for high school English classes probably consume 90 percent of all index cards produced. English teachers seem to think that the only way to research a paper is to use 3×5 inch index cards. However, there are many ways to collect data. Some writers use index cards, others use notebooks, tape recorders, or whatever means best suits their needs. If you have no preferred method, try to experiment and find the method that is most comfortable for you.

Index cards can be useful, and they come in several sizes that may be more convenient if you have a lot of information about a particular topic, but it is difficult to get continuity once you have filled a substantial number with facts. Photocopying the research material and then entering the data into a word processor can be extremely efficient. You can use individual pages like index cards, but gain the tremendous advantage of being able to move data around within the word processor. The result is that you can arrange all the facts in a suitable order and then massage the data into reasonable prose and add the necessary continuity to make a coherent argument.

One technique that allows you to get data into a final form quickly (for example, when a deadline has crept up on you) is to read the reference material that you have collected, and then enter it into the word processor in digested form. Not only does this procedure shorten the writing process, it makes plagiarizing more remote. Since you do not have the source present, and have probably read several different sources anyway, you almost cannot help putting the topic into your own words.

LOCATING SOURCES

If you construct your outline first, the topics that you want to research are clear and well-defined. However, where to go to get the answers to your questions is often not as clear. The public library is the first place that comes to mind, and should be included on your research list. The public library is often not the best source of information, especially for technical topics. People, rather than libraries are frequently the best source of information.

Libraries —Public and Technical

The first place to begin looking for material is the local public library. If you have not been in a library lately, and particularly if you have not been in a public library with a particular research goal in mind, you might be surprised at how much you have forgotten and how much has changed. Today, besides the basic services that are reviewed in this section, many public libraries offer access to on-line databases. Those, however, will be discussed separately.

Getting Started

Major metropolitan newspapers are indexed, and most large libraries will have *The New York Times Index*, which comes out monthly and covers topics ranging from current events— including the paper's excellent Tuesday *Science* section—to recipes. In addition, there are specialized indexes that may be far more useful to you than a generalized index. These include:

- *The Engineering Index* (monthly, annual) provides brief annotations of articles.
- *Applied Science and Technology Index* for industrial and trade topics, including an annotated description of articles
- *Computer Literature Index* (Applied Computer Research Inc., Phoenix, AZ). Quarterly
- *Electrical and Electronics Abstracts* and *Computer and Control Abstracts* (Institution of Electrical Engineers in arrangement with the Institute of Electrical and Electronics Engineers (IEEE), Parsippany, NJ). Monthly

- *Biological and Agricultural Index* covers publications in fields related to the life sciences and is published 11 times a year.
- *Electrical and Electronics Abstracts*
- *Biological Abstracts.* Twice a month
- *Chemical Abstracts,* includes chemical patents
- *Science Abstracts.* Monthly
- *Science Citation Index.* Quarterly
- *Computer and Control Abstracts*
- *Microcomputer Index*

The last item on this list—keep in mind that this is only a sample list—deserves further mention. It is published in four multi-volume sets annually, with supplements. The four volumes are:

- *The Citation Index.* Lists recent papers that cite an early scholarly work or publication in a technology publication. From this index you can discern which older papers are being referred to by recent authors. This information, of course, may tell you what more recent papers have been published on related topics or themes.
- *Permuterm Subject Index.* Uses key words to identify articles by topic.
- *Corporate Index.* Allows you to determine what has been published by members of a specific organization or institution of interest.
- *Source Index.* Provides complete information on recent journal articles.

In the *Citation Index*, there may be an entry like this one from the 1986 index:

COLLINS, T
  78 IEEE Spectrum 15 48
  Looft FJ IEEE Educat 29 100 86

Interpreted, that means that author T. Collins had a 1978 article in IEEE Spectrum, Volume 15, beginning on page 48. This article was cited in 1986 by F.J. Looft in an article appearing in Volume 29 of the IEEE Transactions on Education, beginning on page 100. To find out more about that article, look in the *Source Index* to find:

LOOFT, FJ
  IEEE Educat   29(2)   100–104   86   2R
  Worcester Polytech Inst. Dept Elect Engn.
  Worcester, MA 01609. USA
  Labonte, RC   Lutz, PC—The NASA
  Space Shuttle—A vehicle for university industry
    educational interaction

This refers to the article by F.J. Looft in the IEEE Transactions of Education in Volume 29, Issue 2, on pages 100-104, which appeared in 1986 with two reference notes cited. The author's address is provided and the names of his co-authors, along with the title of his paper.

Although the system takes some getting used to, there is a key in the front of each volume that interprets the arrangement of the data. The volumes provide a useful and efficient way of finding other papers that may be on the same topic as an article already of interest, whether you work backwards or forwards through the search process. The best way is to begin with a work that you know to be pertinent to your work and then look it up to find a list of current works that cite that work as a starting reference. Then, use the *Source Index* to provide complete bibliographic data on those articles, many of which may be useful. In addition, there is a *Corporate Index* if you are interested in finding out what workers at a particular company, university, or other organization have been writing.

The Card Catalog

The card catalog lists all the books owned by the library, and provides this information in three different ways: author, subject, and title. The author card is alphabetized by the last name of the author or the senior author of each volume. At the end of the section of cards for books written by each author are cards concerning biographies, critical studies of that author's work, and so on. By looking under the particular name, you will find everything relating to the author and the author's work.

Similarly, title cards are useful in locating a book if you already know its exact title. This card is considered the main listing for each book, and all the library's different editions, revisions, and translations of that title will be found here, including those in other library branches or in special collections.

Typically, the subject cards are the ones most useful for research. Subject cards list all of the books on a particular topic. If you were looking for information on transistors, the subject cards with the heading Transistor would contain the title, author's name, and the catalogue number of every book that library has on that topic. In using the subject cards, however, you must be creative. If William Shockley wrote a book called *My Role in Inventing the Transistor* it might be found under the Transistor subject heading. However, if he wrote an autobiography called, *My Life, My Inventions, and My Opinions,* it probably would not be cross-referenced in a useful way for your

purposes. It would not even be surprising if there were no subject heading at all for a subject such as *Transistors*. In that case,
you should consider looking under *Computers* and hope there is
a separate subheading for components. Or try *Semiconductors*,
Electronics, or even *Technology*. The title cards are also organized
alphabetically. Generally, but not always, all three types of
card—author, subject, and title—are intermixed.

In many cases, the listings in the card catalogue that are of
most interest are reference works. These are often listed with
the letters "REF," which indicate that those volumes are in a
separate section for reference and do not circulate. That is,
they cannot be checked out and are available for use only
within the library.

Readers' Guide

The *Readers' Guide to Periodical Literature* is often more useful
than the card catalog because information found in magazines
is more current and focused than that in books.

The *Readers' Guide* is a series of annual volumes aug-
mented with monthly and quarterly supplements. The quarterly supplements are a compilation of the three previous
months' listings. Each indexes major consumer magazines by
topic, including feature articles, major news stories, and book
and film reviews. The list of periodicals that are indexed appears in the first few pages of each volume, and most libraries
mark the ones that they carry. Most of the magazines indexed
are readily available.

A typical page from the *Readers' Guide* is shown in Figure 1.
Listings are compiled exactly as for the card catalog but only by
author and subject, not by title.

You will notice on the excerpt from the *Readers' Guide*,
subjects are listed flush left. Amplifications of the subjects are

Footwear—*Continued*
Putting your best foot forward [hunting and fishing] Outdoor Life 169:42 Je '82
The scoop on boots. Seventeen 41:24 S '82
The shock absorbers [exercise sandals as cure-alls for sore feet] D. Blumenthal. il N Y Times Mag p70 My 23 '82
Shoes are made for walkin'. il Curr Health 2 8:10-11 N '81
A step-by-step guide to buying footwear. M. E. Barrett. il Mademoiselle 88:105-6 Ag '82
Sport the right pair. il Seventeen 41:128 Ap '82
Think your shoes are shrinking? There may be another explanation. J. Desy. il Ms 10:93-4+ Mr '82
White bucks [men's shoes] J. Berendt. il Esquire 98:32 S '82
Who lusts after celebrity shoes? These are the kinds that buy men's soles (and women's too). K. Cassill. il People Wkly 17:34-5 F 8 '82
Who says sneakers are bad for your feet? A. Soorikian. McCalls 109:37 Mr '82

Anecdotes, facetiae, satire, etc.
My mother explains it all to you [sneakers] D. K. Mano. Natl Rev 34:1628+ D 24 '82

Care
See also
Shoeshine business
Caring for leather and suede boots. il Seventeen 41:28 O '82
Caring for your boots. il Teen 26:77 O '82

Cleaning
How to clean canvas totes, beach bags and espadrilles. il Glamour 80:41 Jl '82

Exhibitions
These boots are made for painting [Shoeboxes exhibition at White Columns gallery in New York City] Vasari. il Art News 81:14+ D '82

Manufacture
See Shoe industry

Repairing
See also
Shoe repair stores
Shoe Goo guru Lyman Van Vliet cures tattered tennis toes with sheer stick-to-itiveness [inventor of a rubbery sole replacement for sports shoes] S. Koris. il por People Wkly 18:82 Ag 9 '82
Footwear, Running. See Running—Equipment
Footwear, Tennis. See Tennis—Equipment
FOR. See Fellowship of Reconciliation
For colored girls who have considered suicide/ When the rainbow is enuf [television program] See Television program reviews—Single works
For the record [television program] See Television program reviews—Single works
For today's black woman [television program] See Television program reviews—Single works
Forage plants
See also
Alfalfa
Legumes
Soybeans
Teosinte
Good forage management holds feed costs down. J. R. Borcherding. il Success Farm 80:D1 Je/Jl '82
Foraging patterns. Marine fauna. See Marine fauna—Food and feeding
Foraker, Jay J.
The San Antonio Symphony: some problems solved. il por High Fidel 32:MA32-MA33 Mr '82

Forbes, John
Helping the small woodlot owner. il Blair Kechums Ctry J 9:80-3 Ja '82
Forbes, Julia Todd
Pick lettuce all year! il Mother Earth News 73:86 Ja/F '82
Forbes, Malcolm Stevenson
Fact and comment. See issues of Forbes
about
And what's Malcolm Forbes worth? Forbes 130:170 S 13 '82 •
Capitalism and quality [interview] Forbes 130:18 Jl 19 '82 •
Malcolm Forbes. A. Lubow. il pors People Wkly 18:48+ Jl 19 '82 •
Forbes, Malcolm Stevenson, Jr
[Column] See issues of Forbes
Forbes, Mark
Add a distinctive extension phone ring to your telephone. il Pop Electron 20:56 Ap '82
Forbes (Malcolm Stevenson) Collection. See Art—Collectors and collecting
Forbes (Periodical)
Flashbacks. [ed. by D. A. Saunders] See issues of Forbes
Forbes four hundred. See Rich
Forbidden Broadway [musical] See Musicals, revues, etc.—Reviews—Single works
Force and energy
See also
Pressure
Quantum theory
Forced labor
See also
Convict labor
Arabian nights in Miami [mistreatment of servants charges against Prince Turki of Saudi Arabia] M. Reese. Newsweek 99:36 Ap 26 '82
Father Roussel's dancing girls [Belgian priest charged with forcing girls into slave labor] P. Lewis. por Macleans 95:42 Mr 22 '82
Slavery American style [cases of involuntary servitude among migrant workers] B. Jacobs. il Progressive 46:18 Ja '82
Forced landings. See Airplanes—Landing
Forché, Carolyn, and Gomez, Leonel
The military's web of corruption. il Nation 235:391-3 O 23 '82
Forcing (Plants)
Force winter blooms from springtime bulbs. il South Living 17:86+ N '82
Ford, Andrew T.
(jt auth) See Chait, Richard P. and Ford, Andrew T.
Ford, Barbara
Killers of their own kind. il Natl Wildl 20:16-19 Je/Jl '82
Ford, Betty
A Christmas wish. il Ladies Home J 99:61 D '82
Ford, Brian J.
Found: the lost treasure of Anton Van Leeuwenhoek [with biographical sketch] il Sci Dig 90:8, 88-90+ Mr '82
Ford, Catherine
Ill-tempered through the park. por Macleans 95:8 Je 21 '82
Ford, Daniel
C'est la vie—c'est la gastronomie! il Skiing 35:189-90 N '82
Ford, Daniel F.
A reporter at large. New Yorker 58:107-8+ O 25; 45-52+ N 1 '82
Ford, Edsel Bryant, II
Edsel Ford. J. Greenwalt. il pors People Wkly 17:51+ F 8 '82 •
Edsel II looks to the future. W. B. Furlong. il por Saturday Evening Post 254:50-3+ S '82 •
Profile: Edsel Bryant Ford II. T. Hogg. por Road Track 33:22+ Ja '82 •
Ford, Eileen
Eileen Ford: how I find those fabulous faces... P. Battelle. il por Good Housekeep 194:143-5+ Je '82 •

Figure 1

A page exerpted from *The Readers' Guide to Periodical Literature*. Used with permission of the H. W. Wilson Company.

centered. For example, *Footwear* is expanded with items such as *Care, Cleaning, Exhibitions, Manufacture,* and so on.

Author's listings are alphabetized by the author's last name. For example, in the listing under *Forbes, Mark,* the first line is the title of the article ("Add a distinctive extension phone ring to your telephone"), "il" means that the article is illustrated, and "Pop Electron 20:56 Ap '82" means that the article of that title appeared in *Popular Electronics* magazine, volume 20, beginning and ending on page 56, in the April 1982 edition. If the article had run more than a single page, all pages would have been listed.

The Library's Research Department

The most underutilized tool in the public library may be its own research department. The research staff knows what material the library contains and how to get at it. They also know what the library does not have and may be able to tell you how to access that, too, through such procedures as interlibrary loan services. I have used the research department at several libraries, and every time the staff has been eager to help and really enjoyed what they were doing. You can save a lot of time by using the capabilities of the research staff and the expertise they have to offer.

Technical Libraries

There are many technical libraries around the country, usually—but not necessarily—affiliated with a college or university. These libraries specialize in technically related material, and have a much wider and deeper range of periodicals, indexes, and reference works than public libraries. In addition to those found at educational institutions, there are other technical libraries which are administered and sponsored by engineering societies (such as IEEE, ASME and others), and other private organizations. For example, the United Engineering Societies Library at 345 E. 46 St. in New York City covers the needs of a number of engineering societies and their members. It is open to the public at no charge.

While technical libraries have a wealth of information that is not otherwise available, access to the information may be restricted and often such material does not circulate and can be used only at the library. A university library may allow its facilities to be used only by students at that school. However, often no check is made for casual reference use by an outsider. And for more extensive use, arrangements can often be made— including check-out privileges—for a fee.

Before traveling a long distance to use a specialized technical library, be sure you have called ahead to determine their hours of operation (these may change dramatically depending on whether school is in session) and whether you will be able to have access to the information you need. For that matter, a phone call to the reference librarian may let you know if they have the facilities you need. You may be able to gain all the information you need from the phone call, without a trip to the library.

Some large companies also maintain technical libraries. If your company does not have a library, you may be able to use the one at another company in your area. Again, a phone call in advance can save a lot of time and trouble.

ON-LINE DATABASES

Computer networks have made retrieval of research data easy. With a personal computer and a modem, you can tie into one of the many available networks and retrieve entire files of data. There are both large-scale commercial networks and smaller-scale private networks.

The two largest commercial networks are *The Source* and *CompuServe*. You can buy membership on these networks at most computer stores. While they are not simply research tools (they have games, bulletin boards, and interactive communications programs), they do offer a good deal of on-line data, including news, stocks, financial data, dictionaries, encyclopedias, and more.

These two networks are set up to appeal to the mass market. The result is that the user fees are fairly low, but the data is quite general. To get more specific and technical data, you have to go to one of the smaller-scale private networks.

Private Networks

In the age of technology and electronics, many indexes and publications have gone on-line for computer access. Not only is that a way to keep up to date better than a quarterly or even monthly index allows, it is efficient, though hardly as inexpensive. Research of an on-line database requires substantial skills for the novice.

If you are working on a very technical topic, that requires a great deal of research, you may want to enlist an expert or two to help you find the best databases for your purpose. You might be able to find help at your local university library or technical library, but often even those librarians lack the experience or facilities to help you conduct a search, though they may be able to provide a referral service to a qualified researcher who will want to be paid not only for the time spent but the telecommunications charges and the time actually spent on-line.

For example, in Los Angeles, although the California Institute of Technology and the University of Southern California are both important schools, if you want help with an on-line search, you have to check with the University of California (UCLA), where the Engineering and Math Sciences Library has the kind of staff to assist you with just about any topic.

Wasted time is diminished if you know exactly what you are looking for and supply a list of key words and synonyms that can be used in the search. UCLA has an excellent library that could help you on a database search.

At UCLA, you are asked to fill out a form that includes the search topic, and room for a detailed description of the topic, in which you are told, "Be specific and define terms which have special meanings. Include synonyms for these terms. Also, if there are points NOT to be included, please state them." The form also asks what languages you want searched, since databases may contain articles that are not published in English, and there's no sense getting a host of citations that may literally be Greek to you.

You also need to identify the years of interest. And to help the researchers further, you are asked to list some known relevant journal articles published more than six months previously (so there is some certainty they will turn up in the search) and less than ten years ago. Such a list also means you will not be charged for finding things you already know.

Besides the information already indicated, the researcher needs to determine which databases will be searched. Determining that usually involves talking with the searcher to identify the best places to look for the information needed.

The results of such a search can be arranged by title, author, journal name, chronologically, or in other ways, depending what you specify and who the researcher is. You may ask for bibliographic citations only, or abstracts when they are available. You can set a limit on the number of citations or the amount of money you are willing to spend. As in real life, the more work that is done and the more results produced, the higher the cost will be. Charges typically depend on the number of databases searched, the time involved, and the number of items found.

A basic service fee is charged in addition to the actual cost of the search, of course. At UCLA—to give you an idea of what is involved—their own students, faculty, and staff, are charged a service fee of $10 and others are charged a service fee of $35.

Having said all of that, it is important to add that once you have obtained the information needed for the search, you will probably want to continue the research on your own. In that case, you need to know that there are a number of organizations that offer access to multiple databases through a single computer network. Often such services will refer you to qualified researchers in your vicinity. Any of them will be happy to tell you about their facilities and fee structures. These may include special starter rates, and there is a good deal of hand-holding available, not only through customer service representatives who can be telephoned for advice, but through reference manuals and a wealth of on-screen menus and HELP facilities. You can either get the resulting data on-screen or arrange for a hardcopy printout.

The leading multi-database research libraries of this type include the following:

Dialog Information Services Inc.
3460 Hillview Avenue
Palo Alto, CA 94303
(415) 858-3785

BRS Information Technologies Inc.
1200 Route 7
Latham, NY 12110
(518) 783-1161

STN International
c/o Computer Abstracts Service
2540 Olentangy River Road
Post Office Box 30122
Columbus, OH 44172
(614) 421-3600

Dial 1-800-555-1212 to obtain the toll-free number for each that can be used in your area.

Dialog is a Lockheed subsidiary that offers access to some 300 databases—a massive amount of research capability, BRS and STN are both much smaller, but with special properties that may make them more attractive. BRS, for example, is particularly strong in medical data, though its offerings go far beyond that. STN's name stands for Scientific and Technical Information Network. It is a not-for-profit organization that provides a satellite link between the Chemical Abstracts database, going back to 1970, with data from a technical organization based in Karlsruhe, Germany and another in Japan. This international search facility is billed only as far as Columbus, OH—they provide all the networking and telecommunication links after that.

INDUSTRY EXPERTS

Because technical writing generally takes place at the leading edge of technology, there are often few, if any, previously published works on a given subject. In fact, your article may become an item that will be researched by anyone who comes along later with an interest in the same subject. In such a case, industry experts—engineers and scientists working directly with the subject of interest, or professional researchers who keep abreast of the latest developments—necessarily become your chief resource for material. Even if there is published work, industry experts can provide excellent insight and add credibility to your article.

With the growth of personal computers, methods of accessing the information that industry experts have is quite simple—if you have a modem.

Locating the Experts

Determining exactly who the experts are, and making contact with them is a bit more difficult than simply traveling to a library. The best place to start, of course, is within your own company. Researchers, who may be your co-workers, often have valuable information that you may not have imagined. Just mentioning that you are writing an article for a magazine is usually enough to get as much information as you wanted—if not more!

A good place to start looking for experts within your own company is to contact the head of the engineering or marketing department, depending on the type of information you need. These people, usually at the vice president level, not only know what their employees are currently working on, but what projects they worked on in the past that might have yielded useful information for your purpose.

Outside Sources

There are several ways to locate experts outside of your own company. Authors of books and magazine articles are good sources. Even if they are especially prolific authors, they are likely to be flattered to have someone call about their article and recognize them as an expert. They will probably be happy to clarify a point that you have not understood, give you new information that they have, or provide other suggestions and resources, including additional contacts to add to your network.

Articles in most trade magazines and books give the the author's name, company, and the company's location. You can call the author at work, however, a letter may be a better procedure. This is courteous and allows the contact person time to organize materials he or she may be willing to share with you. Remember to treat others with the same respect for their time as you would like for yours. Otherwise you are asking him to invest in the price of a telephone call, possibly long distance, for the privilege of assisting you. With a letter, enclosing a stamped, addressed envelope is almost as much of a requirement as it is a courtesy. Even if the author can afford the postage to respond to thousands of well-wishers each year, why should he have to? Paying your own way is the surest sign of sincerity. If the author's location is not given in the article, you can call the editorial department of the publication, and they will provide a way to get in touch.

Some magazines, both trade and consumer, publish interviews with leading scientists, engineers, consultants, analysts, and industry experts. Such interviews not only are a good source of current information, they also provide another contact name for research. Since the people interviewed are often top corporate management with only limited time, if you do call them for information, be sure you are asking something that could not as easily be found out elsewhere, have your questions written out ahead of time, and keep the call short and to the point. Once again, unless there is some urgent time constraint, sending a letter might well be a better procedure. If you do call one of these people, bear this in mind and have only one or a few short, succinct questions.

The experts within your own company can provide you with names of other experts, as well as making a direct contribution. They make contact with their external peers at meetings and trade shows, which compose another good source of expertise.

The engineering societies (IEEE, ASME, ASCE, ASChE, ACM, and others) have regular meetings, which frequently cover state-of-the-art subjects. These might also be an ideal place to learn the latest developments in a particular field of interest.

The American National Standards Institute (ANSI) sponsors standards committees for virtually every technical and they can put you in touch with the committee of interest. You can write ANSI at 1430 Broadway, New York, NY 10018 and ask to be put in touch with the committee of interest. Most ANSI committees publish a membership roster that lists the corporate members and the name of their representatives.

Research Firms

Research firms are *the* most important source of expert opinion, advice, and data. These companies' entire business revolves around collecting and maintaining every possible piece of information concerning their field of expertise. Such firms then sell their services to industrial customers that need this detailed information to reach major financial and marketing decisions and help in planning long-term corporate strategies.

In addition, such research firms can truly be considered as a technical resource and will often answer most any question you can pose, as long as you credit them with providing the information.

I have frequently used Dataquest, a San Jose, CA, market research firm that specializes in computers. They have individual experts for each segment of the computer industry, including personal computers, data storage devices, and printers. The quantity and quality of their information is astounding. Like a library research staff, the staff of most research firms love to hunt down the answer to your question.

WHERE TO BEGIN

As you have no doubt learned from this chapter, there are a number of places to go if you need help with research. Still the old standby, and the best place to start is the library, either public or private (or university). New methods of research have surfaced in recent years too, and these are often more appropriate for high-tech subjects.

On-line databases are easy enough to access from your home, yet have perhaps the most in-depth technological base of information amasses anywhere. Industry experts and research firms can also provide you with an incredible array of data. You will have to pay for both database access as well as a session with an industry expert or research firm, the amount depending on the facility and the services you need.

Do not let all the options keep you away from trying different research services. The best way is to dive right in. As the Munchkins said in *The Wizard of Oz*, "It's always best to start at the beginning."

S E C T I O N
TWO

WRITING
FOR TECHNICAL
MAGAZINES

CHAPTER
3

WRITING ARTICLES FOR TECHNICAL MAGAZINES

*"Keeping your writing within the
constraints of the specifications
developed will let you write with
more direction and will save you
a great deal of time."*

Why write an article for a trade or consumer magazine anyway? It certainly takes a lot of time, and even if the publication pays for your article, the dollars per hour will be much lower than your normal salary.

One of the easiest ways to get exposure for your company's products, and technology, as well as to promote your company's—and your own—expertise is to write an article for a magazine. You may also be writing about a significant discovery or a new or improved method of doing some process. In these cases, your purpose is as much to advance the state-of-the-art as it is to promote corporate public relations.

Let's all be honest. There is not a single author who minds having his ego bolstered by seeing his or her name appear in print. The first time is the greatest thrill, but the thrill is still there time after time. This often provides the necessary motivational push to get the article done.

Depending on your product and who your customers are, your article might be aimed at a trade magazine or a consumer magazine. The majority of technical articles are written for trade publications. This is because there are actually few technical consumer magazines and because consumer magazines use freelancers and unsolicited material less frequently.

Trade magazines go out to industry executives, marketing managers, and engineers. Be sure you know who the audience is before you begin. One important aspect of writing is tailoring your article to the level of the person reading it. It is a good idea to get an author's guide and audience profile from the marketing manager of the magazine before completing your outline. If a significant portion of your audience is in sales but you do not know it, there is no way for you to capture their interest with some relevant aspect of your topic. We'll talk more about this later in the chapter.

PACKAGING SPECIFICATIONS

You have your outline in good shape; you have done your research; you know why you want to write this article; so you are ready to write . . . right? Well, not quite yet. If you were designing a computer, you would have to define more than just the performance and capabilities of the system. You would also have to define other constraints, such as: it must mount in a 19-inch rack, the floppy disk must face forward, power consumption must be less than 100 watts, and it can't weigh more than 45 pounds. Writing an article is no different; you have to define the packaging constraints before you can start the actual writing.

Whether or not you are writing for a pre-determined or target publication, there is a simple principle that will be useful to guide your work and to present it: Provide as much assistance to the editor as possible. By helping the editor, you are helping the reader. An important overall guideline: State things as simply and directly as possible.

For example, any nonstandard technical terms should be spelled out the first time they are used, not abbreviated. If you introduce a term, identify it clearly. In some industries, such as the computer and peripherals area, terminology changes so rapidly that annual lists of the latest buzz words are published so workers even within the same company can keep track of the new vocabulary that is created to match the new technology that's being developed. Be sure you are up-to-date on the terminology of the area you are writing in.

In order to finish planning the article, you need to adequately answer these questions:

- What audience are you writing for?
- How long should the article be?

- How many and what type of figures are needed?
- What is your deadline?
- When will your article be published?
  (or maybe more appropriately, when do you *need* it published?)
- How do you submit the manuscript?

Let's look at these questions more closely. What audience is the article written for? This is a very important question. You may be aiming your article at the consumer, a marketing professional, an engineer, a combination of two, or all three. Even within these groups you need to define the audience further. There is no sense writing a highly technical discussion of disk drive head flying heights or petrochemical distillation if the readers are all going to be middle managers with degrees in finance.

If your audience is engineering and technical readers, just what is their level of technical competence? An article on computer software that is written for a software engineer should be written very technically, employing the language (i.e., jargon) that the engineer feels comfortable with. On the other hand, if you are writing the same software article, but it will be read by software salesmen or even end users, you will have to write at a different level. You can still write technically, but you will have to explain and define many terms that would be common knowledge to the software engineer.

Once you have determined the audience that you want to read your work, you have to find out which magazine or magazines reaches this audience. This usually isn't too difficult because you probably subscribe to one or more of these publications. A good library will have an assortment of magazines for you to review as well.

Once you decide which magazine best suits your article, you are ready to contact that publication.

There are several reasons for approaching an editor as soon as possible after you have decided to write an article for a particular magazine. The most important reason is to make sure that an article such as you have outlined is wanted. If they reply that they just covered that topic or that they already have someone writing the very piece you had in mind, you can start thinking about where else such a piece could be placed. Another possible response will be to provide you with a deadline so that your article can be used in a specific issue. The editor may have planned a special issue on a relevant topic or have some other reason for wanting to make specific plans to use your manuscript in a particular way. An editor is grateful for timely queries from competent authors who want to write the very article needed to fill a given editorial slot and which would otherwise have had to be solicited.

You also need to consider the scheduling and production time associated with publishing an article. For you to receive the June 1 issue on or before June 1, the articles have to be complete and ready to typeset and lay out about six to eight weeks prior to the publication date. You also have to allow for editing, sending the galley proof to the author, returning them to the publication, and proofreading. About four to eight months will pass between the time you submit your manuscript and the time the magazine containing it reaches its readers.

Which brings me to another and very practical reason for making an editorial inquiry before you write rather than wait until you are finished: Someone else may write your article first and get it in to the magazine ahead of yours. And if the editor tells you that the piece you propose is not what he needs just now, he may have some other ideas that would be appropriate for his needs and still be topics you would be qualified and able to address. It has often happened that a writer has called me to pitch one topic and has ended up writing several articles, none of which, perhaps, is what he or she originally had in mind.

While the telephone is the most efficient way of establishing a link with a publication, it is not always the best way. Editors, in spite of abundant evidence to the contrary, are human too, and have more to do than time permits. Taking a phone call from a stranger to discuss an article is not always possible. A better plan is to look in the magazine you are targeting and find the name of the editor who seems most appropriate. Then send a brief query to that person by name. Typically, the title you are looking for is the New Products Editor, Features Editor, Technical Editor, Senior Editor, or Managing Editor. Match the position to your article's viewpoint. By all means, include a notation such as "Manuscript Proposal Enclosed" on the outside of the envelope. Trade publications get literally hundreds of pieces of editorial mail each day, and you want to differentiate your proposal from the press release announcing XYZ Company's new vice president.

Include with your query as much of your outline as you have. A sketchy proposal may be rejected because you fail to give the editor enough information to make a good decision. When you write to the publication, try to keep your letter to a single page. Begin by stating the topic of your proposed article, and then provide a brief description. Explain why the topic would be appropriate for the publication that you are querying, and conclude with a bit of information about yourself and why you are qualified to write this article. Remember that the magazine's editorial space is a limited resource, and the editor must choose only the articles and authors that he thinks will maintain his standard of quality, and allow the publication to continue selling advertisements.

Another reason for contacting the magazine is that while they may not have a need for exactly what you are proposing, they may need something similar that you would want to do. The response might be, "No, that's not what we'd like for a feature. We wouldn't be able to use it for six months or more. If you could make it into a two-page box, however, I could use it with an article in two months."

Like product design, you must not only have a good idea, but a customer who is willing to buy the product. There are a lot of great ideas that made it to a product, but no customer was willing to buy.

Figure 1 shows a sample of a letter to a publication that meets the requirements just stated, with a minimum of bother to you. Note that there is a definite timetable for followup action so you don't have to wait forever for a reply. If there is no response by the time mentioned, you should phone or write a second letter of inquiry. You should expect some response within six weeks of submission.

Multiple submissions of the same article to different publications are almost always frowned upon. There are instances when you might want to do this, for example, if it has critical time value. If you do, be sure to advise each publication of what you have done and what your intentions are (offer to the first response or whatever). The more acceptable way to handle this circumstance is to follow the procedure outlined in the previous paragraph, and if you get a negative response or no response, submit it to another appropriate magazine. If you are doing so because of a lack of a response, be sure to send a letter to the first publication informing them that you are submitting your manuscript elsewhere. (Use tact, you may want them to consider another one of your articles later.)

If you have decided to write an article that would appeal to the audience of some magazine that you do not receive, or perhaps do not even know exists, you have a more difficult situation. In this case, you will find it most practical to find the names of some magazines in the right field and decide which one sounds most appropriate. Sounds like another trip to the library is in order.

416 Fets Gate
Automata, CA 95999
December 1, 1988

Mr. Ed A. Torr
Technical Magazine
30 Offset Drive
Masthead, MN 46858

Dear Mr. Torr:

As you are well aware, there has been a great deal of activity in the field of superconductivity of late. Of particular interest is the application of superconductivity to transportation. At Magnetomotive Corporation, we have been engaged in researching superconducting traction motors.

Clearly, this would have a significant impact on the transportation industry. I would be interested in writing a feature length article for your publication, dealing heavily with the problems and possible solutions associated with superconducting traction motors. Attached with this letter is a brief outline.

I have written five technical articles, published in such publications as Electric Motor Design, Traction Power Weekly, and High Current Review. I also hold three patents for related work.

Since this is a timely article, I would hope to have it published within the next five months. Therefore, I will follow-up with you right after the new year if I haven't heard back by that time.

I hope that you will find this a stimulating topic, and I'm looking forward to hearing from you.

Sincerely,

Al Lectron

Figure 1

A typical cover letter that you might submit to a publication, along with an outline. Note that it describes your article, its importance, your qualifications, and it also establishes a timetable for follow-up.

The names and a brief description of magazines can be found in annual publications such as *Ayers Directory of Magazines,* and *Standard Rate and Data.* Most libraries carry both of these, and as always the librarian will be happy to help you locate them. Unfortunately, knowing a publication's name and the name of its editor is not enough. If you have never seen a copy of a publication, it is doubtful that you can address its audience successfully.

Another publication which can help with consumer magazines, but not with technical material, is *The Writer's Marketplace.* I will always remember the publisher of a small magazine filled with ghost stories and supernatural horror à la Stephen King and his description of the magazine's editorial content. He wrote, "If you don't normally read this type of fiction, you won't be able to write it."

Looking at a comprehensive list of magazine titles in your field is a good start, but it is only a start. You have to look at an issue or two of the most appropriate magazines to see whether they reach the audience you are trying to address. Fortunately, you can find most of these publications in technical libraries. If not, call the magazines that seem most appropriate.

Tell them frankly, "I have an article I think you'd be interested in, but I'd like to see a copy of your guidelines for authors before proceeding. I'd also appreciate seeing a copy of a recent issue if you can provide one." You will almost always get a copy free of charge.

So far, we have been discussing the initial research for finding a way to reach your intended audience. On occasion, you might be contacted by an editor to write a specific article. In that case, the author's guide and a copy of the publication should be provided by the editor in question, the public relations agency, or the company officials who have asked you to do this article. If these are not provided, ask. If that does not work, insist!

How Long Should the Article Be?

The next parameter for the article that needs to be specified is length. This is best determined after you have spoken with the targeted publication. The editor will often give you this information immediately. Sometimes, however, there is some flexibility. If you have already written the piece, and it does not match what the magazine wants, you may be able to negotiate the length with them. The safest bet, though, is to get this spec directly from the editor.

Before getting too deeply into length, let's consider some basic rules for manuscript preparation. First, manuscripts are *always* double spaced, typewritten (or mechanically printed) and on one side of white paper. Margins should be about 1½ inches top, bottom, right, and left. There are no exceptions to these rules.

A page of a manuscript prepared in this way contains roughly 200 words. This is an important conversion, because editors frequently specify length as the number of words. Using a word processor that counts words is helpful for this task, too.

A feature-length article, that is about four magazine pages, consists of about 3000 words of text (15 manuscript pages) plus three to six illustrations. Shorter articles and one page boxes or sidebars are proportionally shorter.

Once you have agreed to a length, stick to it, or discuss any change as early as possible with the editor. If you have agreed to 3000 words, 2900 will be ok, as will 3100, but 1500 will come as a major shock if that is all you deliver. If you asked a machinist for a nine-inch long, inch square piece of brass, you would probably be quite angry if he came in the week before your prototype is due for a project review meeting and said "I could only get one five inches long. They were on backorder until next month so I figured this would do." Filling four magazine pages with a 1500 word article is quite a trick!

Do not try to cover up the discovery that you only have half as much to say about a topic as you had thought you did. A good writer can expand something simple to endless lengths as needed, all the while adding words, but not adding information. My old boss had a poster in his office that said "Never confuse motion with action." In this case, "Never confuse words with information."

How Many and What Type of Figures Are Needed?

This part of the spec is very closely related to length in some ways. The actual length of the article in magazine pages is the summation of the text and the figures. However, the length specified by the editor is for *words* and must be met in addition to supplying illustration material. At the time the editor gives you the word-length requirement you will also get the requirement for figures.

One common excuse for delivering an article much shorter than it is supposed to be is that extra figures were provided instead. Substituting illustrations for words just does not work. Try asking a cook whether you can add more sugar to a cake recipe to make up for a shortage of flour! If you want to substitute figures for text, talk with the editor.

Just as bad is supplying too many figures, on which the article is dependent. If the requirement was for four to six illustrations, and you submit 12 that build upon each other and are not easily cut, you have again put the editor into a tough situation.

There is more to the figure requirement than just quantity. The types of figures must also be considered. Type refers not only to the way it's produced, but the content as well. Most publications do not require camera-ready art. That is, it does not have to be ready to put directly onto a page of the magazine.

Nonprofessionally drawn sketches of sufficient clarity and detail are suitable. You should always supply appropriate labels for parts of the figure. Also supply captions for the figures, written in full sentences and conveying *information!*

The degree of detail and technical content of the figures is related to the article itself, and the requirements of the publication. When I select illustrations for an article, I pick the ones that are able to describe a principle better with a picture than with words. Since space is a premium, the converse of this is the best way to synthesize figures.

In addition to drawings and computer printouts, you may also want to use photographs. Black and white photographs should be clear, in focus and have good contrast. The size of black and white prints should be 5×7 or 8×10 inch (or something in between).

Color photographs are preferred in most instances. Color photographs are published using an entirely different technology as black and white photos. Because of this, transparencies (or slides) are preferred to prints. 35 mm slides are acceptable, but larger sizes such as 60×60 mm through 60×70 mm (both on 120 size film) reproduce with unbeatable quality. If you simply do not have access to anything but a print, go ahead and submit it. It can be used, but it is more time consuming and expensive to reproduce.

What Is Your Deadline?

The deadline can be obtained only through contact with the publication. You will discuss it with the editor and reach a date that is mutually satisfying. Often intermediate deadlines will be given. They may give a separate deadline for the outline (usually only a couple of weeks after the proposal) and for the manuscript. This allows the editor to better measure the chance of the article getting completed—and unfortunately quite a few proposals never make it beyond that stage.

Missing a deadline can be everything from catastrophic to having little consequence. Unless you have strong reason to believe otherwise, however, respond as though missing the deadline would be catastrophic. Editors sometimes need to fill holes left by other missed commitments, and have given you the absolute last date that your submission can be prepared for publication. By all means, contact the magazine as soon as you know if you are going to have trouble meeting the date.

When Will Your Article Be Published?

As mentioned at the beginning of the chapter, this question might more appropriately be stated as "When do I need the article to be published." Let us discuss these questions.

There are several variables that affect when a particular article will be published. Magazines are always scheduled several months in advance, so you will almost always have a lag time of four to eight months between when you submit the final manuscript and when the article is in print. In addition to this "manuscript pipeline," your article may be held to be part of a feature or issue focused on the subject about which you have written about. Sometimes your article is bumped due to lack of space, and is rescheduled for a later date.

Sometimes, the article slips into a crack. All things considered, it may be even a year or more before the article is published. I have seen, especially with consumer magazines, 18 months pass before one of my articles was published. Unfortunately, these variables on rare occasions may result in a previously accepted manuscript being rejected later.

With this background, let's take into consideration when *you* need to have the article published. Often, one of the purposes of writing the article was to have it published in conjunction with another event—a product introduction or trade show, for example. At other times, the actual publication date, within reason, is not critical.

In cases where you desire the article to be published within a certain time frame, discuss the problem with the editor. Usually, you can arrive at a plan that meets both your requirements and the publication's. Bear in mind the discussion of scheduling variables and do not expect a blanket commitment such as "I'll absolutely run this in the February edition." This sometimes happens, but is by no means the rule. If you are unable to reach an acceptable publication estimate, it may be time to find a different publication to fit your requirements.

HOW DO YOU SUBMIT THE MANUSCRIPT?

The procedure for submitting the manuscript will come from the editor. Different magazines have different submission procedures, and with the growing use of personal computers and word processors, the methods and formats are somewhat varied. In general, you will find that most publications adhere, at least roughly, to the following guidelines.

Regardless of what electronic means for submitting the manuscript may be acceptable, you will *always* need to submit a hard copy—that is, a printout of the material on the disk. Some publications require two copies, and whatever you do make one for yourself and keep it! Manuscripts do get lost, sometimes in transit and sometimes at the publisher, but it does happen. If you are a fan of Murphy's Law, you probably know that your manuscript will only be lost when you do not have a copy, or when the only copy you have is on that disk that just got destroyed.

The manuscript should be typed (or printed with a computer and printer), double spaced, and only on the front of white paper. Margins should be roughly 1½ inches on every side. Heard this before? It bears repeating; you'd be surprised how many single-spaced manuscripts are submitted by authors who "absolutely understand the requirements" or "have written many times before."

Electronic Submission

Many magazines are set up to accept some form of electronic submission of the textual material. This relieves the enormous data input problem that magazine publishers have. Consequently, more and more publications are encouraging this method of submission. Electronic submission can take one of several forms. It is possible that a publication would want the manuscript on tape, but that is really doubtful. Most likely they will want it on a floppy disk, via a modem directly to their computer, via a modem and electronic mail, or via a facsimile machine.

For submission on a floppy disk, you must determine the format and word processing program to use, if any, that the publication can accept. Almost universally, the disk format will be IBM PC compatible. The choice of word processor will not be so universal. Although there is about as much emotion regarding word processor choice as there is for automobile choice, the major word processing programs that most publications will have access to are WordStar™, Word Perfect™, and Multimate™. Others (good ones at that) exist and may be usable if the magazine that you are working with has a copy of the program. If you are not able to match word processors, they can give you alternate ways to submit your manuscript on disk, such as via an ASCII file. Some publications prefer an ASCII file to any word processor version.

Many publishing firms have on-line modems that allow you to transfer your data directly into their computer. Sometimes, the computer is actually the magazine's typesetter's computer. You should discuss and completely understand the magazine's requirements before attempting direct modem-to-modem contact.

The third method of electronic submission is via an electronic mail service. This method is nearly as popular as floppy disk. By far, the most popular electronic mail service is MCI Mail. With electronic mail, the submission and retrieval is totally at the user's convenience. If you finish at midnight, you can call the electronic mail modem and dump your manuscript into the magazine's mailbox. Then the editors can retrieve your mail at their convenience.

Finally, there is the facsimile, or FAX, machine. These machines have been around for many years, but seem to have become popular only in the last decade or so. A FAX machine sends copies of pages directly, using the telephone as the channel. By using FAX, you can electronically transmit not only the text, but figures and tables as well.

Always get the name of your contact and address the manuscript directly to that person. Write on the outside of the envelope or package "Manuscript enclosed." Your manuscript has a much smaller chance of being misplaced if so labeled.

Protect artwork and photographs properly for shipment. Make and keep a copy of each piece of art as well as the manuscript. Pack photos between pieces of rigid packing material such as heavy cardboard. I don't know how much attention is paid by transportation companies (U.S. Postal Service or private carriers) to the precautionary legend "Do not bend," but it certainly cannot hurt to write it on the envelope.

A final word on submitting your article involves permission to use certain materials. This most often applies to artwork. It is never a bad idea to get a signed release from any recognizable person in a photograph. This may or may not be required, but is easier to do up front than trying to run down the subject after the magazine contacts you and requests one. If you are using artwork that your company has generated, ask the publications manager or marketing vice president if it is okay to use it for the purpose you intend. I have seen some very upset corporate executives because of failure to do this. Get a credit or source line in writing from the source and use on your artwork without fail.

A WORD ABOUT WORD PROCESSORS

Comparing word processors to typewriters is like comparing car travel to walking. You will probably get there, but it is a lot more time consuming and trouble, and you may run out of desire before you do! In short, if you do not have access to a word processor, figure out a way to get access or buy one. By far and away, the most common use of personal computers is for word processing. I use my computer for word processing more than 97 percent of the time. The balance is spent programming and using spreadsheets or "playing around."

Word processors allow you to correct mistakes easily, rearrange text effortlessly, copy sections of text from one file to another, and check your spelling. Accessory programs even check grammar and style.

Buying a word processor program is a personal decision. I prefer WordStar and have used it since 1982. WordStar is more difficult to learn than some of the more recent programs, but it is very powerful and widely used. What amazes me the most about WordStar is that it has all of the functionally useful features of the very newest programs. WordStar was one of the first, and established many of the baselines for word processor programs.

The people who do not like WordStar generally had trouble learning to use it. Once learned, most people swear by it (and sometimes at it!). This is the area that more recent word processors have improved upon . . . ease of learning.

What Can a Word Processor Do?

Most of the work that you will use a word processor for relates to the typing of your manuscript. Unlike a typewriter, you can instantly correct mistakes and rearrange text by hitting a few keys. Figure 2 is an example of editing and entering text with a typical word processor. This example is from WordStar, but most word processors are similar.

In addition to the typing function, other parts of the word processing program give you access to power never possible with a typewriter. Most programs include, or have optionally available, a spelling checker. WordStar's spelling checker, CorrectStar, is shown in Figure 3. Not only does the spelling checker find misspelled words, but it offers suggested corrections as well. If a word is not included in the CorrectStar dictionary, you can add the correct spelling of the word to your personal dictionary so misspellings will be caught in the future.

Many word processors now offer an on-line thesaurus to help you find synonyms to vary your presentation and increase your vocabulary as well. An example of WordStar's WordFinder is shown in Figure 4. Sometimes, not all of the synonyms suggested by the thesaurus are actually synonyms in the context you are using the word. In these instances, you have to pick a word that has the correct meaning. In Figure 4, for example, "voyage" or "flight" would not be appropriate synonyms in the context illustrated.

Another use is that word processors are handy for making

Figure 2
This display from MicroPro WordStar is typical of what the computer pre-
sents to the user with a word processing program. Courtesy of MicroPro
International.

forms for frequent use. In Figure 5, MicroSoft's Word has
been used to create an expense form. The capability of making
horizontal and vertical borders makes this task easy. You can
also use the bordering feature as shown in Figure 6 for making
tables. Tables such as this can be included within your technical article.

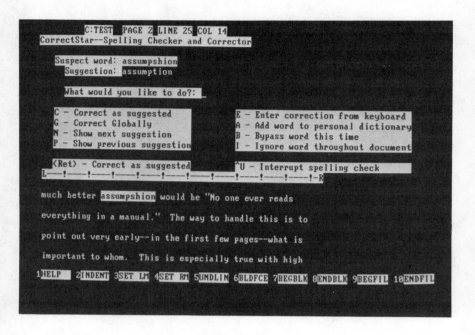

Figure 3
CorrectStar is typical of many spelling checker programs that complement
word processors. Misspellings are caught and corrections are suggested. Courtesy of MicroPro International.

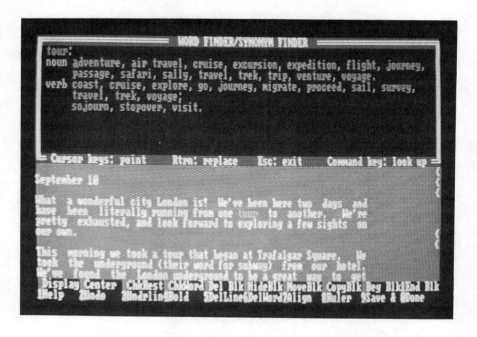

Figure 4
WordFinder is an on-line thesaurus that can provide you with synonyms
where requested. Courtesy of MicroPro International.

Name:	Harry Watkins					
Purpose of trip:	Presentation to National Accounts					
Date of trip:	May 23, 1987		Date submitted: June 1, 1987			
Enter mo/day	Mon.	Tues.	Wed.	Thurs.	Fri.	Total
Hotel	$456.78					$456.78
Airfare						0
Taxi		6.45				$6.45
Tips			2.34			$2.34
Meals						0
Entertainment						
Total	$456.78	$6.45	$2.34			$465.57

Figure 5
Word processors, such as Microsoft's Word, can also be used to create commonly used business forms such as this expense report. Courtesy of Microsoft Corp.

```
┌──────────────────────────────────────────────────┐
│ THE MACFARLAND MANUFACTURING COMPANY              │
│ TWO BLANKENSHIP LANE                              │
│ FORT MITCHELL, KENTUCKY 40234                     │
└──────────────────────────────────────────────────┘

          Summary of Financial Results
     for the fiscal year ended January 31, 1987
     ─────────────────────────────────────────

We are very pleased to show strong gains in revenue and
profitability over last year:
```

	1986	1987	Change
Sales.............	$98,000	$151,000	54.1%
COGS..............	44,100	60,400	37.0%
Gross Margin......	53,900	90,600	68.1%
Fixed Expenses....	9,900	11,900	20.2%
Profit before taxes	$44,000	$78,700	78.9%

Figure 6

Horizontal and vertical bordering, available on many word processors, such as Word, enables you to make tables along with your manuscript. Courtesy of Microsoft Corp.

Selecting a Word Processing Program

Criteria for selecting a word processor varies somewhat, but for writing technical articles, the following is a good starting point:

- It should be no more difficult to learn than the time you are willing to invest.
- You should be able to scroll forward and back within your text, and see previous and subsequent pages.

- The program should respond fast enough for your needs.

- You should be able to move blocks of text within and outside of your file.

- There should be a compatible spelling checker program available.

- An on-line thesaurus can be as helpful as the spelling program.

- Finally, the word processing program needs to be popular enough that you can send floppies and expect the recipient to be able to read them, or it should be capable of creating an ASCII file.

SUMMARY

Once you have established your preliminary specification with your outline and have done your background research, you need to do your packaging spec. This includes selling the idea to a prospective magazine, and acquiring the specifications or requirements that particular magazine has for the preparation of your material.

The packaging specs for your technical article will include the following parameters:

- Level of audience
- Length of text
- Type and quantity of figures
- Submission deadline

- Publishing date
- Submission procedure

Keeping your writing within the constraints of the specifications that we have developed will let you write with more direction and will save you a great deal of time. It will hopefully save you some rejections as well!

CHAPTER
4

PRESENTING YOUR IDEAS

"Presentation of your ideas is an important step to seeing your article in print."

Now, we are ready to begin looking at getting your ideas down on paper. Chapters four and five will deal with this subject. This chapter deals with the form and style—sort of a system architecture. Chapter 5 covers the mechanics of writing—the design rules for writing.

No matter how good your idea is and how thoroughly you have researched the topic, it will not be as easily accepted by your targeted publication if it is poorly prepared. We've seen some of the criteria that affect the presentation of your manuscript. Now, let's delve into the issues of style and form.

WRITE AS IF YOU WERE IN DISCUSSION WITH A COLLEAGUE

When you are discussing a technical topic with a colleague, the conversation is very efficient—you are just talking. You would never say "when one compresses the spring . . .," you would say "when you compress the spring . . . ." The use of the impersonal you has become perfectly acceptable for technical writing. This takes a good deal of the stuffiness out of articles. All that formality made the articles unduly dry and ensured that a substantial number of readers would never finish reading the piece. In short, use a *conversational style*—engage your reader.

English teachers have frightened many potential writers away from becoming published writers. In many ways, English classes are like engineering classes. They teach you the fundamentals, and more importantly, they teach you a way to think. Just as you would never calculate the transfer function for every design, nor run through a stress analysis for designs that you have done many times, you bend some of the rules of English when you write.

Formal English training is essential, but communicating is more than using formal English. While proper grammar, spelling, matching of subject and verb, and the other language rules are important, of greater importance when writing a technical article is conveying an understanding of the subject, to the audience.

For proof that a conversational tone is better at communicating than stuffy, formal English, pick out an article that you really enjoyed, and that you felt communicated the author's point to you clearly. Almost invariably the article was written, and read by you, as if a direct conversation between you and the author was taking place.

Some of my friends have complimented my writing, saying "you write like you talk." This really pleases me, because that has been, and is my chief objective. Not only that, but it's by far the easiest way for me to write . . . I just talk to the keyboard.

You can take this approach too far. It does cloud the understanding to say such things as "the transmission line sees an infinite impedance" or "connect this guy to this other thing . . . ." A particular danger is the attempt to ingratiate yourself to the reader by attempting humor, humorous analogies, and funny or smart-aleck phrasing. This false joviality does not communicate information effectively. Furthermore, what is humorous when spoken is not necessarily humorous when read.

Be Careful with Technical Terms

In using the conversational style, be careful not to inject too many acronyms and jargon without sufficiently defining these terms. If you are using a new term, or an old term that doesn't get much usage, be very careful to adequately define it. More familiar terms such as CPU, psi, and AWG need not be defined, but be sure that what you see as familiar is really familiar to your reader. You have to make a judgment call, but it is best to be a little conservative and define all the borderline terms. This practice provides assistance to the editor, as well as the reader.

Nonstandard technical terms should be spelled out the first time they are used, rather than using the abbreviations right from the start. Computer buzz words change so frequently that even experts can have difficulty keeping abreast of current terms and jargon.

HOW TO PRESENT THE MATERIAL FROM YOUR OUTLINE

You have your outline, so you know just about what will go into your article. The amount of each topic that you want to include in your article is still open for consideration. There are several ways to present the same outline. Some will be successful, and some will not.

In considering how to present the information in your outline in finished form, remember to keep in mind the targeted reader. Instead of trying to imagine thousands of faceless engineers, doctors, marketers or other readers, think of one in particular . . . yourself. What would *you* want to know if you were reading this article to learn about the topic? What would a colleague need to know if you were explaining the matter to him? Again, the emphasis is on conversational style.

Two possibilities may be useful when determining the content mix from your outline. If you are dealing with a new technology, you can pretend you are writing a letter or having a conversation directly with a colleague. You can envision the colleague as a new hire, whom you are trying to bring up to speed on the subject. Since the reader will be familiar with the general areas of the subject and the basic vocabulary involved, once you refresh his memory with a general introduction to the main issues, you can proceed to zero in on the precise topics of importance.

If, on the other hand, you are to explain something well-known to you and not particularly new, you might pretend you are explaining it to an intelligent graduate student at a university. What do you know about this subject that someone coming in for the first time is likely to need to know? The trick in this instance is to provide links from what the reader may know when he begins, to what you want him to know when he finishes. That may require a brief review of some relevant fundamentals to serve as groundwork before you elaborate the areas of interest. The use of analogies is a very useful vehicle for this primer type of article.

The most common mistake when writing a technical primer is to take too much knowledge for granted and to underexplain rather than overexplain. In reviewing your outline before actual writing, look at each topic and try to find the assumption that you make about it. It may not be valid to believe those assumptions can be made for the reader, or that he has the same preparation and background that you do.

A good example of this from everyday life is in the sports segment of the radio or television news. "They've taken up playing Billy Ball," or "Kareem throws up a brick," have little or no meaning if you don't know who Billy Martin is or that a brick is a very bad basketball shot attempt. Now, those may not be unfair assumptions for a sports report, but if a statement similar to that is central to communicating an important technical aspect of your article, you better be certain that you do the extra hand-holding to be sure the meaning is clear to the reader.

Sentences like the one about Billy Martin operate as bottlenecks to the flow of information because access to so much of an argument can be eliminated if they fail to communicate properly. In other words, if you do not understand the first sentence, the rest of the point will likely be missed or misunderstood.

One time the Vice President of Engineering called a meeting of the entire design staff, of which I was a member. It was somewhat of an ass-chewing meeting, but he turned it into a motivational sermon. He finally reached the pinnacle of his speech and punctuated it with "Gentlemen, it's time to fish or cut bait."

Now, there is not a great deal of commercial fishing going on in Central Indiana, and the engineers were all rather young (under 30). Instead of everyone jumping up and shouting "Yeah, let's go get 'em!," 16 engineers stared blankly at each other. We had all missed his point because he did not communicate with us in terms that we could understand.

LEAD THE READER THROUGH YOUR ARTICLE

Beyond looking for the embedded assumptions that may need to be clearly defined before a topic or subtopic in your outline can be expanded upon, explained, and clarified, there are some ways to help the flow of your article and conduct the reader along without missing any vital connections. The three important ways of doing this are with connecting words and sentences, use of good lead sentences, and through a generous sprinkling of subheadings throughout your manuscript.

One way to make sure the connection between each thought is made clearly is to ask yourself what the connection is between each pair of connecting sentences. Perhaps it is possible to begin one sentence with therefore, however, or, or another connecting word. Such words not only link important materials together, like the way letters are linked together when you write, they also help the flow of ideas continue in the right direction. By using such words, you get a paragraph that is not just a set of unconnected assertions.

Another help to the reader is to begin the article or paper with an explicit statement of what you are going to explain, and as importantly, why. In a consumer magazine such as you would find on the newsstand, articles often begin by backing into their subject in a way designed to be interesting and to lure the reader onward. In this approach, bits of information are dangled in front of the reader as if they were bait for a fish. These are effective techniques and are often used in fiction writing: "Little did I know the great surprise in store for me when I opened the . . ." Having caught your interest, such articles then go on with all manner of background explanation, perhaps not getting to the lure again until near the end: "Oh yes, about the surprise . . ."

Typically, a technical article has more important purposes than to provide amusement for the reader or to while away his idle hours. If the readers had idle hours, they would not need you to help pass the time! Consequently, it is a good idea to begin the article by announcing exactly what is going to be in it, but to do so with an interesting lead sentence. The opening paragraph or two may well amount to an abstract or precis of the whole article. You may even wish to say explicitly, "This article will discuss . . . ," although there are better ways of stating the purpose than to slap the reader over the head with it.

There are several advantages to this approach, whether you employ the direct approach such as "this article is about . . . ," or a more subtle method. It may interest the reader if you write,

"Everyone wondered who killed Colonel Mustard" and delay revealing the murderer until the last paragraph. However technical articles require more respect for the reader's time. A better approach is to say, "Colonel Mustard was killed in the library with the candlestick by Professor Plum!" Even if you do not add, "This article will explain the motives for the crime and demonstrate how the truth eventually came out," the reader will have decided whether he wants the details. This type of lead is called a *summary lead.*

One important advantage of the summary lead is that it helps the author to focus his thoughts. If you find that you cannot summarize clearly what it is that you intend to say, you will probably find that you are trying to tell a story that is poorly focused, or which is not particularly coherent. Not only does that kind of summary lead allow the reader to turn elsewhere if he is not interested, it has the added advantage of telling him everything you wanted him to know anyway, so by the time he decides to turn the page, if that is his choice, he has already learned the essence of what you intended to communicate.

You can use summary leads throughout the body of the article as well as in the lead paragraph. If you look at Chapter 3, you will notice that the entire chapter was summarized in the first few paragraphs. Then, as each topic was discussed, the entire gist of the topic was summarized in its own lead.

A summary lead that reveals all, before the reader can decline the author's invitation, is particularly useful in new product stories. Incidentally, in articles of that type, it is especially important to stick with the facts and not make assertions that the editor or publisher may not find easy to believe. You should avoid writing product or technology descriptions that include words like first, fastest, simplest, best, and most efficient. Unless you can present hard data to back up these assertations, there are likely to be unpleasant letters to the editor if the editor is so foolish as to allow such unsub-stantiated or controversial claims to get into print in the first place. This kind of hype often turns up in press releases, and isn't out of place in product brochures, but it rarely makes it to the page of a magazine and your best bet is not to try it. If your product is fantastic, it goes without saying . . . so don't.

One of the old axioms of writing is to tell the reader what you are going to say, say what you want to say, and then tell the reader what you just said. Summarizing like this can also be done several times in the course of an article by means of summary statements, but it must also be done at the end to round out and summarize the article in a way that provides a feeling of completion or closure. Often, writers forget (1) to provide an introduction to the material, (2) to provide any solid overview of what is to come, or (3) to put their abstract at the end. If you find this is the most comfortable way to work, go right ahead. Just do not forget to move your conclusion to the front of the manuscript when you are done, and write a new ending to replace it.

Use Subheads

The third way to help the reader follow the thread of your discussion is to use subheads to announce new topics. These may be topics that have already been itemized on your outline, or they may be portions of an itemized topic that has turned out to require more words than you had expected it to take. Either way, the important point about subheads is that they are a help to the reader and must not be integral to the structure of your story. That is, the article should still make sense if all the subheads are omitted. Here are a couple of examples. The "wrong" way:

Clock Frequency
 is not a good measure of CPU speed because instructions can require different numbers of cycles.

The right way . . .

Measuring CPU Speed
 Clock frequency is not a good measure of CPU speed because instructions can require different numbers of cycles.

Subheads are often left out during the process of editing, layout, and printing, so they cannot be used for such purposes as a change of topic. Also, readers engrossed in the narrative often skip the subheads as they might in a newspaper article or the way chapter titles in an exciting novel are often overlooked as an interruption to the flow of the story. If you want to change the topic, the subhead is not the way to do it. Instead, you need to find a way to do it directly in the main body of your text:

- Before looking at the implications, it is necessary to review the fundamental causes . . .

- This brings up another problem with thermo-magneto-optical . . .

- At the same time as you are evaluating these efficiency issues, you must also evaluate several other factors of equal importance . . .

- With this as a background, let's look at new developments in plating technology . . .

Since subheads may be omitted, it is also necessary for you to avoid making important announcements in the subheads that the reader needs to know about. For instance, you cannot list five alternatives for discussion and put each one into a subhead, then say something like, "This alternative is the one that's most interesting." If that subhead gets deleted, the reader will never know "What alternative!?" By all means, use the subhead to announce the topic, but then go on to repeat the information in the text.

"Bullets" can be used to list items that you want to discuss. Bullets are solid dots which delineate topics much as the roman numerals do in an outline. Do not go overboard here though; using too many bullets dilutes the effect of the ones that *are* important.

OTHER TECHNIQUES

The presentation of ideas is going to be made first to the editor, then to the readers. Thus it is helpful for you to provide a headline of some sort so the editor can have an idea at a glance what you think is the subject of the manuscript. Each magazine or publication has its own style of headline, but to introduce your topic to your first and most important reader, that doesn't matter. You do not need to write a headline that fits some predetermined style; that is something the editor is skilled at doing, and even if you spent the time to provide exactly the right length and format, the editor would almost certainly re-write it on general principles.

What you can do is help focus the editor's attention by saying clearly and succinctly what the article is about. Use a verb, and write a complete sentence. Do not just say "New topics in carbon bonding," say "Carbon bonding techniques are finding a wide range of applications in exothermic reactions." Do not just say, "Techniques for increasing data density," say "Lower head flying heights and vertical recording increase data density." The label raises a question that it does not answer. In this case, the question is, "What about the techniques for increasing data density?" That is, it makes the reader wonder what you are going to say about them, and does not tell him whether you are going to say anything he wants to know.

FIGURES AND TABLES

The difference between a table and a figure is how the data in each is arrayed. A table has numbers or data arranged in regular columns and frequently can be set into type by the typesetter. A figure often contains artwork, derived from data, that must be drawn by an artist. A list of topics would be a table while a schematic would be a figure. A pie chart or a bar chart would be a figure, while the data to construct these figures would be a table.

Unless camera-ready art is specifically requested, and it very seldom is, figures and tables may be prepared on the computer or word processor, or may even be hand-drawn. What is important is that they be large enough to see and have clear labels on all necessary parts. It is helpful to provide both a title and a caption for the illustration, although in many cases, tables will not use the caption and the figure may lose its title to match the publication's own style standard. Nonetheless, it removes a burden of determining the figure's purpose from the editor and helps avoid the possibility of errors creeping into your work if you provide the caption and title both.

Another thing to remember about the artwork is that if it is to be redrawn, it must be clear enough for the artist to do so correctly. Technical artists often do brilliant work that is not properly appreciated, and may turn typical engineering napkin scribbles into polished presentation graphics, but that doesn't mean the artist is an engineer who knows what those peculiar abbreviations on the labels are intended to suggest.

Be sure all parts are labeled correctly. Use as few abbreviations as possible, especially if the abbreviation is something that is not spelled out in the text or caption and is not something that could easily and certainly be located in a reference work. There is no harm in marking a distance as "5 in."—especially since "in." can be looked up readily. However, if your meaning has any possibility of being misinterpreted, write out the label in full. It may not fit elegantly into the space provided, but it may still save a long distance phone call from someone asking what you meant. Once again, data can be removed or terms abbreviated, but the editor cannot manufacture data that you did not supply.

For figures and tables both, be sure to supply captions in clear English sentences, complete with verbs. Do not say, "System Block Diagram," as that is completely obvious. Nor should you make that into a sentence and say "This is the system block diagram." Those are words, but they don't convey much information. What your caption should say is, "Here is the block diagram of the system. Note how the buffer isolates the input from the output, allowing the slow disk to be used with the fast CPU." Now that says it all—and it took very little work on your part since you know its purpose intimately.

This seems pretty trivial, but you can find endless examples of this type of caption in the "Proceedings" that are published for many trade shows and symposiums. The reader may ask "What about the block diagram?" or "Why is this here anyway?" Your job is to obviate these questions before the reader has a chance to ask them.

The purpose of figures is to help tell your story. So go ahead and provide an extra bit of handholding in the caption and summarize what the figure is supposed to reveal. Do not just say it is a timing diagram, translate it to explain why it is there and why it is important to the reader. The reader can probably figure out the answers on his own, but why should he have to?

For presentation, all the figures should be grouped at the end of the article, with one to a page. Never intertwine the figures with the text. Captions should also be numbered and grouped in one place, but on a single page. The captions should have the same number as the figure (they do not always end up that way if you do not double check them!).

Thus you are supplying a manuscript, captions, and figures and tables as three separate items. Since neither one is any good without the others, they must also be supplied at the same time. The capabilities of today's desktop publishing are great for company newsletters, but cause problems if you try to do too much of the editor's and publisher's work for them.

The way that the figures are tied into the text is via references. Never leave out any figure references. This makes the editor take a guess as to where you intended the figure to go, and his guess doesn't always correspond to where you thought they should go. Figures are referred to by consecutive reference numbers. For example:

"The gear train (Fig. 3) transfers the . . ."

"Figure 4 shows the correct way to . . ."

"The radius at the top (Fig. 5) is significantly wider than the radius at the bottom (Fig. 6)."

If a figure logically has more than one part, refer to the parts as a, b, c; such as Fig. 4a and 4b.

Number Everything

Finally be sure that you number everything you submit. It does not matter whether you number figures and tables together or separately—the editor will decide what is appropriate for his publication. What does matter is that everything have a number so that it can be referred to readily. Each page of the manuscript must have a page number as well.

Visuals help the reader, as subheads do. If you find that your article doesn't seem to have any appropriate illustrations, then think some more until you come up with a few. Certainly there are times when far more are required, and times when visuals are hard to imagine. Roughly speaking, one graphic element for every 3–4 pages of manuscript is about right, depending on circumstances.

If you have doubts about what is appropriate, talk to the editor. Ask whether he wants photographs, can use color or black and white, or requires camera-ready art. Most of all, if there are to be no graphic elements, or if they are to be delayed for any reason, let the editor know as soon as possible.

If all this emphasis on keeping the editor informed or meeting the editor's needs seems excessive, the reason is not that I have been an editor and know what frustrates me, and what fails to work. It is because the editor truly is your customer. He is the one whose requirements must be met if your work is to be presented properly and accomplish the communications tasks you intend. If he does not know what you are trying to say, doesn't see the relevance of the graphics, is left to make up his own captions, and generally does not understand what you are trying to explain, then it is fairly certain his readers will not understand either. That means the article will be rejected, sent back for a rewrite, or will be "fixed" inhouse in a way that may be neither accurate nor appropriate as far as you are concerned. In other words, if it comes out at all, it may come out wrong or severely delayed.

DEALING WITH REJECTION

If you do enough writing, you will have pieces rejected. It happens even to the best writers. Every writer knows that some of his work will be rejected. No one goes unrejected forever.

You can minimize your chances for rejection, or stated in a more positive way, you can maximize your chance for acceptance by being certain that you follow the specifications exactly, and don't try to "sneak in something."

As an editor, I have had to reject many articles. I felt somewhat bad about most of the ones I rejected, because it was clear that the author had put a lot of work into the article, but he had not followed the specifications that I had laid down for him. I have to admit however, I felt very good rejecting some of the pieces! Let's look more closely at why articles get rejected.

Follow the Editor's Directions

When you make a proposal to an editor, you generally arrive at a compromise between what he needs to serve his readers, and what you wanted to write. Usually, this is a fairly minor compromise, with the author's suggestions serving as the guidelines. However, since the editor must meet his reader's needs, his salesmen's demands, and his available space, you will run the risk of rejection if you do not follow his guidelines.

The most frequent reasons for rejection are: Incorrect length, wrong technical level, and not paying attention to the publications particular policies.

Length

Length was discussed earlier, so I will just mention it briefly here. If an editor is anticipating a 20-page manuscript, and you send him only 10 pages, there is a good chance he will not be able to use your article at all. Even if your 10-page article is fantastic, he has a certain number of pages to fill, and might be forced to fill them by using an inferior article that *is* the correct length.

On the other hand, if the editor has allotted space for your 10-page gem, and you overwhelm him with 30 pages of fantastic prose, you will end up in the same boat. More is *not* better; for that matter less is not either. "Just as specified" is what the editor is expecting, and your best chance of having your article accepted is by getting the length just right.

If you run into a length problem, whether too short or too long, get in touch with the editor immediately and discuss your problem. By warning them early, you might be able to work out the space problem.

Wrong Technical Level

This was discussed earlier, but bears yet another mention because it so frequently results in rejections of otherwise good ideas. By reading a copy or two of the magazine that you are trying to get published in, as well as talking with the editorial people, you will be able to determine the technical level at which you need to write.

My magazine runs very technical descriptions of new products each month. When the articles are recruited, the authors are given a more or less standard spiel:

We need a 3- to 5-page, double-spaced technical paper about your product. It should describe, in technical detail, how the product works, what problems it solves, what possible solutions you tried, why the solution you chose was the best, and any other technical breakthroughs. Please don't include any marketing "hype," and be sure to double space your manuscript.

Shortly after making this speech to an author, his manuscript arrived on my desk, describing his new computer terminal. Not only was there not a single technical word in the article, but one of the paragraphs (describing the "technical" merit) read: "This new terminal additionally features a choice of cabinet colors that rivals the latest in aerobic wear . . ."

Obviously, this article was rejected. Clearly, the author had not met the specification that I had given. His technical level was way below that of our audience, so we did not have a need for his article.

More common is writing too technically for your audience. Since you are very familiar with your topic, it is easy to forget that it may have taken you months or years to understand what you expect your reader to grasp in 30 minutes of reading. Think about how you arrived at your understanding and make sure you provide enough details for your colleagues to understand your topic.

Most trade magazines are technical, so your articles will naturally be technical in nature. Where the "overtechnical" problem most often develops is when you are writing for a journal that covers a discipline different from your own. An article on the use of conductive plastics for electronics that would appear in an electronic magazine could run into this problem, for example. The electronic engineer knows electronics, and *something* about plastic, but he is nowhere near the expert that the plastics engineer is.

By the way, you will notice that double spacing is mentioned twice in the recruiting speech. Even so, about 20 percent of the articles come single spaced.

Not Following Publisher's Policy

This may be the most frequent reason for rejecting an otherwise good article. Publications have certain policies for good, albeit complicated reasons. You must follow the policies.

One of the policies of my magazine is that it is a technical journal, not a product magazine. Therefore we discourage mention of the company's product. We do try to allow a passing mention because of the time and effort that the author has put in. However, we want the mention to be an example of applying the technology being written about.

Even with this explicitly stated in the author's guide, the type of articles appearing in past issues, and discussions between the author and the editor, we receive many articles each month that are simply product promotion pieces. These are rejected. They do not meet our basic criteria: No advertisement disguised as an article.

Double spacing is a policy of every magazine. If you are going to write, double space it! Some other common policies are assignment of copyright to the publication, obtaining model releases for photographs with people in them, obtaining permission to use photographs and illustrations provided by a company (even your own), submitting a copy of the article on diskette (and including a printed or hard copy version), and others.

Learn and follow whatever policies the particular publication has.

Unsolicited Articles

The previous discussion assumes that you have spoken with an editor and have agreed on a topic, length and so on. Many articles are written unsolicited, and the only correspondence with the editor is when your completed work lands on his desk.

You need to follow the previous discussions even more carefully to give your manuscript its best chance for getting printed. What you will have to rely on most is study of previous issues of the magazine that you want to write for.

If you are not writing for a trade publication, you should also pick up a copy of *The Writer's Marketplace,* which lists virtually every nontrade magazine and a partial list of their requirements.

Look over back issues. You will find that most articles are about the same length. Try to make yours the same length.

Are the articles written for a constant technical level, or are they varied? If the technical level is constant, write as close to that level as you can. If the articles are varied, write at a level that you think would appeal to the greatest number of readers . . . and is comfortable to you.

Notice any other subtleties you can about the magazine. You would not want to submit a product review to a magazine that does not publish reviews. Try to get a flavor for the types of articles the magazine publishes, and tailor yours to fit. Alternatively, find a magazine that has a format to fit what you want to write.

GETTING FROM THE OUTLINE TO THE MANUSCRIPT

One easy way to begin writing is to expand your outline so each item is turned into a complete sentence with a subject and verb. To make each point completely clear, you may find you need several sentences to explain exactly what that point is supposed to be about. If you cannot write at least one sentence, you must not have anything to say—leave that item out.

In many cases, a sentence you have written to expand a

point on the outline can be considered as the topic sentence for a paragraph. Before expanding further, ask yourself what the connection is between the various items and see if you can find a way to help the reader along by using linking words that make clear what you intend the flow of the narrative to be. Words like *thus, consequently, on the other hand, for example, although,* and *in contrast,* not only help the reader understand, they help to guide you in writing so each idea does have a real relationship to the material that comes before and after.

Otherwise, you may be in the position of simply listing a quantity of facts without showing how they are connected with each other, as if your article were to be titled "27 Ideas about Hydroelectric Power" instead of something specific, like "Low Head Hydro Proves Feasibility at Numerous Sites," in which you would need to develop a number of ideas that would have to be related to one another in a way that made sense.

Transitions integrate the various sections of your article and make it into a unified whole that holds together (is coherent) and moves toward a distinct conclusion. In short, they help you create in the reader a sense that he is being told a story, that there is some sense of pattern or connection among the various elements of what you are writing. Otherwise, each subsection seems to start a new topic from scratch.

Such transition words, and even transition sentences, are especially useful when a new topic is about to be introduced. Since you cannot rely on the use of a subhead to signal this change, you need to announce the new topic in a timely fashion. Readers should not be surprised by the direction of your argument or discussion, and your means of delivery should not be that of a parent conducting children to various exotic exhibits at a zoo, each of which is new and unexpected.

Instead, begin by saying the equivalent of, "As we walk through the zoo, we will see many different exhibits. First, we'll take a tour bus ride around the park to get an idea of its extent.

Then, we'll visit the monkey house before lunch. After lunch, we can see the reptiles." By sketching out the scheme of your article for the reader, you are kept on track and the reader is too. Similarly, it is useful to repeat key elements of your argument from time to time, not simply as a matter of redundancy, but as a reminder. This can be done quite casually by summary statements and can be used to introduce a new topic as well. For example:

- Having seen that inventions are moving out of the lab and into the home with greater speed every year, it is now possible to consider the future of technology.

- While the first generation of plastics were limited in form and function, today's compounds have a much wider range of capabilities.

To make sure the reader is following your line of thought, be sure to define terms as you go. That does not mean that every technical term has to be spelled out, but it does mean that you must be sure not to lose anyone by using a term they do not know. Even an audience of colleagues is rarely as knowledgeable as you assume they are. In the computer industry, for example, engineers who work on video display terminals may not need to be told that a pixel is a picture element, but those who spend their time increasing the capacity of disk drives may not know what the term means unless you tell them.

The casual way to do this is to say "a picture element (pixel)," using the common term first and then the jargon. Equally, you might say something about ". . . pixels. These picture elements . . ." Either way, you are sure everyone is still with you.

Recently a situation came up in an article I was editing where the author had referred to part of a video signal as the "front porch." What is the front porch do you suppose? Is there anything about the term that suggests it is "a blank signal area used for video positioning"? I didn't think so either, and tracking down an engineer who could define the term took a lot more time than I cared to spend. Yet without including the definition, not one reader in a hundred would have known what he was talking about.

Be Specific

A technical article or paper for a technical audience, even more than a general interest article for casual readers, needs to be as specific as possible. Do not just say, "there are many disadvantages," but list them. Do not just say, "a part with too wide an operating temperature cannot be used effectively," say "a part with an operating range of +/−20 degrees F will burn out twice as often as a component with half that operating range." Or whatever may be true. The point is that you should avoid making vague statements that cannot be backed up.

One way to do that is to read each sentence aloud and ask what it means. (This is a great help at the outline-sentence stage, incidentally.) If the answer is that it means what it says, you're in good shape. If it means something else, however, you should find out what that something else is and say that instead.

One of the best ways to be specific is to use the precise word. Mark Twain pointed out that the difference between the right word and the almost correct word was "subtle but enormous" like the difference between lightning and the lightning bug. If a person says something is a "mute point," (silent) even if you know they mean the point is "moot" (arguable) you may think they are a little on the dumb side. The best plan is to be sure you know the meaning of the words you use, and to look up the meaning of any words you are not sure of.

Another way to be specific is to support your idea. Just asserting something, especially if you are making a controversial or novel assertion, is not sufficient to make people believe that what you are saying is true. "I am not a crook" may be 100 percent accurate, but it does not automatically persuade the listener. Ask Richard Nixon. What is needed after such an assertion is support, for instance, in the form of authority, reasoning, demonstration, illustration, or example.

For technical writing, the best form of support is the addition of relevant data in the form of explaining what you mean or describing something. Reference to a figure, chart, graph, table, or other visual aid can be helpful as well. Case histories that demonstrate some principle, product, technology, or procedure in action can make your point vividly and succinctly.

Besides examples, which make your point directly or provide the type of data that led to your conclusion or assertion, a good way to support your points is to use an analogy. Analogies do not provide anything, since what is true for one situation may not be true of another, but they do allow something to be explained in a way that makes use of the reader's own customary terminology. And what it takes to communicate something you know is to put it in terms someone else can handle, recreating your knowledge in his reality. Analogies are a powerful way to do just that.

Here for example, is I. Dal Allan, gently educating by comparing data storage standards to levels of education:

These extended examples are more elaborate analogies than most, and each served as the basis for a whole article. More commonly encountered are much more casual analogies, as when a heart is compared to a pump, a circuit diagram to a map, or a computer to a brain. Analogies make meanings clear by using something known to explain something that may be unknown.

WRAP UP

The general structure for an article should follow a simple plan. Imagine someone—your wife or husband, for example— asking what kind of an article you are going to write, the short answer may be, "I'm going to show how the Jones process worked over at the Smith Company." Or, "I'm going to show how the use of on-line databases has simplified some types of research." Without worrying about putting the ideas in a technical form, you have already summarized your purpose in writing, and thus made clear, at least by implication, the type of structure your article will have.

Typical purposes for writing are:

- To state a problem and provide a solution
- To show why or how something works (a new product, for example)
- To demonstrate a principle, procedure, product, or technology through one or more real world examples (case histories)
- To make a point by building up examples or data

Each type of article contains an inherent structure. Beyond that, all that is really required is that you put down all the necessary information patiently, without hurry, allowing as many words to explain each point as seem to be required. Some points can be made simply by stating them; others require more elaboration and support before they have been made sufficiently well to allow you to move to the next point with confidence.

Letting the reader know what to expect is important. As the opening paragraphs of an article set the stage for all that follows, each section and paragraph can also have a topic sentence that states the point that will be made, elaborated upon, argued, or investigated by the rest of the words in that section or paragraph.

Remember, your primary goal in writing a technical article is not to impress the readers with your vocabulary or mastery of the English language. Rather it is to communicate an idea as clearly as possible. This means using a more conversational style, as if you were talking to a friend. Do not be afraid to use the word "you" in your article.

Help the reader to understand your points by leading him through the material. Use connecting words and sentences. Subheads help in leading the reader through the article, but make the transition between topics within the text, not through subheads.

Include figures and tables where possible and where helpful. Always present them separately from the text. Be sure to number the figures and tables, and by all means refer to every figure within the text. Provide complete and understandable captions for the figures. Presentation of your ideas is an important step to seeing your article in print.

CHAPTER
5

DESIGN RULES FOR WRITING

"This book focuses on how to get your wonderfully structured sentences onto paper, into the proper form, to the editor, and most importantly, published."

You didn't pick up this book to learn proper English, spelling, grammar, punctuation, and sentence structure. Of course, these are important elements of writing, but you already have these functions somewhat mastered. This book focuses on how to get your wonderfully structured sentences onto paper, into the proper form, to the editor, and most importantly, *published*.

This chapter will remind you of the important elements of writing, illustrating many of the points by examples rather than by theory and rules.

I prefer to look at the constraints to writing which are imposed by grammar and "proper" English as the design rules for writing. In designing a bridge, there are design rules that govern the minimum strength steel to use for a given load, the wind factor, rush hour capacity, and so on.

When designing a computer, you must remain within the design rules with respect to voltage, available current, physical size, and so forth. English, grammar, style, spelling, and structure are the writer's analogy to these design rules.

The editor's job is to take a manuscript from an author and polish its structure, grammar, spelling, and style before publishing it. The editor can fix a lot of mistakes in a manuscript, and will, if he wants to publish the idea badly enough. So if you are submitting an article about a world-class idea, don't worry too much about the "English" part of the paper: It will be published. However, if your idea is somewhat less than "world class," you should spend some time brushing up and relearning the proper rules of English, as well as good writing style.

The outline will provide a good, solid foundation. If your outline is well constructed, the logical organization will serve as the medium for your message. If you have ever done any computer programming, you know that while the logic of a program is by far the most important criterion, if you do not have proper syntax for the commands, the program will not work. Even though you have a solid logical base, you have to work on the "syntax" to ensure that your article will be published. An editor can sometimes be so badly turned-off by poor construction, grammar, and spelling, that he will reject the article on these grounds alone.

Most editors *expect* technical writing to be poor. You will be allowed a bit extra slack because of this. However, you still should review and try to use the suggestions from four broad areas: Sentence construction, writing at the proper level, spelling and grammar, and using consistent units. The goal of your writing is to get your message across. Concentrating too heavily on the grammar may interfere. It is a good idea to write your whole manuscript, then in another pass through, once you are sure the concepts are fully covered, review and correct the grammar. Examples of typical problems follow. As you read this section, analyze what your particular writing shortfalls may be. You may also want another person to read your manuscript for proper English.

SENTENCE CONSTRUCTION

During the time I was editor for a magazine, I was amazed by the condition in which some manuscripts arrived. Not everyone retained the basics of grammar, particularly not all "writers"! The basic requirements of a sentence are threefold: It must have a subject, it must have a verb, and it must have punctuation. These rules are taught in grade school. However, a good many people have forgotten these three simple requirements. Let's take a quick look at each.

Omitting the subject from a sentence is the least violated requirement. Even when the subject appears to be omitted, it may be implied, which is within the rules. This is often referred to as the "implied you subject." Here are a couple examples of implied you:

Get out! (You *get out!*)

Please help me get up (You *please help me get up*.)

Total omission of the subject is rare. An example of this is:

Flying over the barn.

A general rule to use to check yourself is to find what the sentence acts, or is acted upon. In the above example, *nothing* is acted on (*what* is flying over the barn?).

The most common problem with sentence construction is omitting verbs. It would seem that this would be difficult to do, but it happens frequently. A sentence must contain an action word (or a verb of being). You can use nearly the same rule of thumb to find the verb of a sentence. I have collected several examples of verbless sentences from actual manuscripts. Here's a sample:

A burning need. (*to do what?*)

Thought waves generated with a force of magnitude never before encountered by her. (*Although 'generated' seems like a verb, in this instance, the phrase "generated with a force of magnitude never before encountered by her" modifies the subject, rather than acts, or acts on, the subject. The thought waves must do something.*)

The large flask containing the liquid. (*How does the flask act?*)

Mr. Smith and three other people. (*Did what?*)

94

None of these sentences has a verb that indicates how the subject acts or is acted upon.

Punctuation is also a problem, especially commas. Sometimes, an author uses too many commas. This problem is easily rectifiable. Worse is when a writer fails to use commas when necessary. It is important to separate clauses and elements in a list with commas to remove any ambiguity of meaning from a sentence. Here is an example of one sentence that can have two drastically different meanings, depending upon position of the comma:

The temperature was raised 80 to 100 degrees.

The temperature was raised 80, to 100 degrees.

In the first sentence, the temperature was raised somewhere in the range of 80 to 100 degrees. The absolute temperature could have been raised from 400 to anywhere between 480 and 500 degrees. In the second example, the temperature was raised from exactly 20 to exactly 100 degrees (80 degrees). The omission of a comma makes quite a difference.

Leaving commas out of a list can also result in ambiguities.

The data is routed to the x, y and z inputs, and the output.

The data is routed to the x, y, and z inputs, and the output.

The first sentence means that there are two inputs for the data: (1) the x and (2) the y and z, which is a single input for both variables. The second sentence implies three inputs: (1) the x, (2) the y, and (3) the z.

Two other problems with commas do not cause ambiguities, but they do result in difficult-to-understand sentences. The first problem occurs by omitting commas that should be used to set off what is really a parenthetical phrase. For example:

The new printers Smith's Press had equipment problems.

The correct form of this sentence is:

The new printers, Smith's Press, had equipment problems.

The second problem occurs when commas are used to tie several simple sentences into one long and difficult to understand sentence. More about this problem under the heading "Simplify Your Sentences."

Very few sentences in technical writing deserve the attention that the exclamation point brings. A good rule is not to use it at all. Perhaps a more practical rule is to use it only when a truly important sentence deserves the exclamation. Some technical editors eliminate exclamation points whenever they appear in a manuscript.

Activate Your Verbs

Active verbs are always preferred over inactive (or passive) verbs. You cannot write every sentence in active form, nor would you want to. Key sentences benefit from action verbs though. And, you make a stronger point with an action verb than with a passive form.

The basic difference between active and passive verbs is that with an active verb, *the subject acts,* whereas with a passive verb, *the subject is acted upon.* Passive verbs have a helping verb (a form of the verb "to be") coupled with them. Many times you can improve your paper by replacing sluggish verbs with more active ones. Here are some examples:

The computer is going to allow us to be more productive.
The computer increases our productivity.

Key sentences can derive a benefit from action verbs.

Key sentences benefit from action verbs.

The answer was calculated by the computer.

The computer calculated the answer.

Replacing these sluggish verbs will make your paper stronger and is especially important in a persuasive paper, where you are attempting to change the reader's mind.

Subject and Verb Agreement

Another all-too-frequent problem is disagreement between the subject and the verb. Singular subjects require singular verbs, and plural subjects require plural verbs. In most cases, this is easy. However in some cases, the subject is not always clear, such as:

John and Tim is the answer.

In this sentence, the subject is answer, *not* John and Tim, so the verb must be singular. This is more apparent if you restructure the sentence to be: The answer is John and Tim.

Simplify Your Sentences

Long and complex sentences are difficult for the reader to understand. Complex sentences require the reader to determine the relationship between the several elements that you have woven into a single sentence. The technique that is often misused is to string several sentences together with commas.

> *Every available unit was called out but it was not enough,
> huge forest fires in the northern part of the state had si-
> phoned away some equipment, brush fires in other parts of
> the county took others, and within hours the fire was com-
> pletely out of control, roaring down the canyons faster than
> a man could run.*

That sentence actually was submitted in a manuscript. The easiest way to eliminate problems such as this is to read your paper
aloud, or in your head. Not even a olympic swimmer could get
through that sentence in one breath. Editing this sentence into a
more understandable thought yields:

> *Every available unit was called out, but it was not enough.
> Huge forest fires in the northern part of the state had si-
> phoned away some equipment and brush fires in parts of the
> county had taken others. Within hours, the fire was com-
> pletely out of control, roaring down the canyons faster than
> a man could run.*

This reconstruction is easier to read and understand, than the
unedited version.

Who Does "It" Belong To?

Another common problem with sentence construction is using
relative or demonstrative pronouns (he, she, it, these, that)
whose antecedents (that is, what the pronoun refers to) are missing. Here's an example:

> *Bill hit John with his hat.*

The ambiguity with this sentence is, with *whose* hat was John
hit with—Bill's or John's? There is no way to tell with this

construction. I have often seen an attempt to repair this sentence so it results in:

> *Bill hit John with his own hat.*

The problem still exists: Did John get hit with *John's* own hat or with *Bill's* own hat?

The cure for this problem is to be sure to adequately qualify the pronoun or not to use a pronoun at all when it's use is unclear. This sentence's meaning is clear:

> *Bill took off his hat, and hit John with it.*

This time, the "it" is clearly referred to within the sentence.

Keep Paragraphs Short

Effective communication is best accomplished with simple, but elegant words, sentences, and paragraphs as well. Writing endless paragraphs are not quite as hard to follow as endless sentences, but do add difficulty. Write short paragraphs, usually no more than a half-dozen sentences or so. Paragraphing gives the reader time to take a quick mental break, and to change gears as you change topics.

WRITING AT THE PROPER LEVEL

I have read a great many articles, written by intelligent people, in which the sole purpose seemed to be to use as many complex words as possible. The articles seem to be saying "I'm intelligent, and to prove it let me show you all the big words I know."

Frankly, these articles do not impress me. It is much more difficult to write an article using simple words and simple, informative sentences.

You should never use a big word when a small one will do. The perfectly ironic phrase is "eschew obfuscation," which means "avoid confusion." Sometimes, one small word won't work, although two small words communicate better than one large word.

A lot of research has been done into reading and comprehension levels. There is an upper limit to people's reading comprehension level, which is roughly equivalent to the highest level of their education. This makes sense, since additional education usually increases one's vocabulary.

Just about everyone prefers to read material that is several levels below their maximum comprehension level. This does not necessarily imply laziness; readers want to expend most of their energy extracting the message rather than working on their reading comprehension level.

Several algorithms have been developed to measure the level of education required to comprehend a given piece of writing. One of the most popular of these is called the Fog Index.* By analyzing a portion of an article or manuscript, you can determine a value corresponding to the number of years of schooling that the reader would need to understand the passage.

The Fog Index is calculated as follows:

1. Select a 100-word passage from the manuscript that you have written. Then divide the total number of words by the total number of sentences to find the average number of words per sentence.

---

*Robert Gunning, *New Guide to More Effective Writing In Business and Industry*, Gunning-Mueller Clear Writing Institute, 1963.

2. In the same 100-word sample, count the number of words containing three or more syllables. Do not count proper nouns, compound words like "time-keeper," nor verbs that became three syllables by adding "es" or "ed."

3. Sum the average number of words per sentence and the number of words containing more than three syllables. Multiply the sum by 0.4 to get the Fog Index.

The first 100 words of this chapter end with the word "designing" in the third paragraph. There are five sentences, so the average number of words per sentence is 20. Eleven words have three or more syllables, resulting in a sum of 31. Thirty-one times 0.4 yields a Fog Index of 12.4. For the average college graduate, the level of reading will be about three to four years below their maximum comprehension level. This results in a comfortable reading level for them.

Try using the Fog Index on something that you have written in the past. Then apply it to a paper that you write after finishing this book. You will probably find that your earlier writing was too complex. You may also verify for yourself that simple writing is not easy. It is also interesting to index articles from several magazines. Publications such as the *IEEE Spectrum* have amazingly high Fog Indexes, while a newspaper such as the *USA Today* will be at a much lower level.

Other indexes include the Flesch Formula, the Fry Graph, and the Flesch-Kincaid Formula. The Flesch Formula and the Fry Graph use graphical methods to arrive at an index. The Flesch-Kincaid Formula is an equation similar to the Fog Index. The Flesch-Kincaid Formula is important because the United States Department of Defense (DOD) requires that contractors producing manuals for the armed forces apply the Flesch-Kincaid Formula to the manuals.

To find the Flesch-Kincaid reading comprehension grade level, follow the following formula:*

F-K grade level = [0.39 × (avg. number of words per sentence)] + [11.8 × (average number of syllables per word)] − 15.59.

The average number of syllables per word is found by counting the number of syllables per 100 words, and then dividing by 100.

These metrics give similar results to the Fog Index. For the first paragraph of this chapter, the Flesch Formula scores "fairly difficult" (on a scale of "very easy" to "very difficult"); both the Fry Graph and the Flesch-Kincaid Formula result in the eleventh grade level.

Software to Aid Writing

Spelling programs, thesauruses, and general grammar programs are available for use on your personal computer. Spelling programs have been available for several years. They help not only the poor speller, but good ones as well. The main advantage of spelling programs is that they catch most typos and they catch words that are habitually misspelled. They do not get all typos because some typos result in other words. I sometimes mistype the word "from" and end up with "form," which the spelling checker will not flag because it *is* spelled correctly. The spelling program is not clairvoyant, so it cannot figure out what you meant, only what you wrote.

Some word processors also have computerized thesauruses. These programs allow you to access an on-line thesaurus,

* Philip J. Klass, "Software Augments Manual Readability," *Aviation Week and Space Technology*, January 11, 1982.

providing you with a list of synonyms for a selected word. These programs have mixed value. On the plus side, you can improve the level of interest of your writing by varying your words, and you can improve your vocabulary. On the minus side, use of a thesaurus may change your writing style somewhat, and unfortunately it sometimes suggests words that are not synonyms within the context of your topic.

Recently, general writing aid programs have been introduced. One example of these, and one that I have used and can recommend is RightWriter by RightSoft of Sarasota, Florida. This program uses artificial intelligence to analyze a document and provide a number of services, including a sentence-by-sentence critique, and a summary of the entire document is given with a series of writing indexes from which to polish your text.

Figure 1 is an example of a passage from a sales letter as analyzed by RightWriter. The sentence-by-sentence critique appears in upper case. Note how artificial intelligence not only determines the problem with a particular sentence, but offers a suggested correction as well.

In Figure 2, the summary of the sales letter is given. The readability index is similar to the Fog Index; in fact, it uses the Flesch-Kincaid Formula. The program calculates the readability index from the entire document rather than from one or a few 100 word samples. This produces a better average grade level than just using a few samples.

The strength index measures the clarity and conciseness of the delivery. The strength index gives high rating to simple, but effective words and sentences written in the active voice. Other factors that the index considers when rating a document are clichés, positive rather than negative tone, slang, abbreviations, and ambiguous phrases.

The descriptive index determines the quality of the way that you use adjectives and adverbs. The values are skewed toward

```
The current status of dealer sales is terrible. Our traditional
    <<* 13. REDUNDANT. REPLACE current status BY status *>>
dealer sales strategy does not seem to be working. Current sales
    <<* 11. AMBIGUOUS?: traditional dealer sales strategy *>>
figures seem to be indicative of a major problem.  In view of the
    <<* 44. USE VERB FORM. REPLACE be indicative BY indicate *>>
fact that the entire company's future depends on dealer sales, this
    <<* 2. WORDY. REPLACE In view of the fact that BY since *>>
problem must be acted upon expeditiously to develop a multiplicity
                    ^<<* 21. PASSIVE VOICE: be acted *>>
    <<* 7. REPLACE expeditiously BY SIMPLER quickly *>>
of solutions to overcome the problem (e.g. better point of sale
    <<* 33. PARENTHESIS NOT CLOSED *>>^
material, more market research, more support by reps, etc.
                        <<* 31.  COMPLEX SENTENCE *>>^
                    <<* 39.  CAN SIMPLER TERMS BE USED? *>>^
                <<* 17.  LONG SENTENCE: 45 WORDS *>>^
```

Figure 1

RightWriter uses artificial intelligence to determine faults in sentence construction and offer suggested improvements. Courtesy of RightSoft, Inc.

good technical writing, which uses enough modifiers to illustrate the message, but not so many that the writing becomes wordy. Lengthy, flowing descriptions should be limited to your novel writing.

Technical manuals must have some jargon for the reader to feel comfortable. However, the level of the reader's understanding dictates the amount of jargon that you can employ. A computer software manual that may be used by first-time users should have less jargon than a manual written for a professional programmer. A supplement to the program allows you to set thresholds for jargon so RightWriter does not unnecessarily flag jargon that is important to the document.

The power of artificial intelligence enables the program to analyze sentence structure within the rules of English. The program actually looks for patterns within the writing rather than looking at individual sentences. Heeding the messages

```
                    <<** SUMMARY **>>

      OVERALL CRITIQUE FOR: test

      READABILITY INDEX: 11.39
   Readers need an 11th grade level of education to understand.

        Total Number of Words in Document:  75
        Total Number of Words within Sentences:  75
        Total Number of Sentences:   4
        Total Number of Syllables: 125

      STRENGTH INDEX: 0.00
   The writing can be made more direct by using:
                        - the active voice
                        - shorter sentences

      DESCRIPTIVE INDEX: 0.61
   The use of adjectives and adverbs is in the normal range.

      JARGON INDEX: 0.42
   The writing contains a good deal of jargon.

    SENTENCE STRUCTURE RECOMMENDATIONS:
              14. Consider using more predicate verbs.


                    << WORDS TO REVIEW >>
   Review the following list for negative words (N), colloquial
   words (C), jargon (J), misspellings (?), misused words (?),
   or words which your reader may not understand (?).
     EXPEDITIOUSLY(J)  1                    INDICATIVE(?)  1
      MULTIPLICITY(J)  1                         REPS(?)  1
         TERRIBLE(N)   1
              << END OF WORDS TO REVIEW LIST >>
```

Figure 2

RightWriter's summary lists several indexes from which you can improve your writing. Courtesy of RightSoft, Inc.

will add variety to your writing and make it more interesting to read.

RightWriter doesn't have a spelling checker in the sense that a word processor does, but it does list words that are questionable. You can then check them to see if they are spelled properly. The checker will also point out jargon and misused words so that you can correct them as well.

From the word frequency list, a list of all words used in the document and a tally of the number of times each word is used is generated. Running this feature takes quite a while on lengthy documents, so you may not want to use it. However, the program does list the total number of words in a document as well as the number of different words. This can be a handy feature if you are writing for a particular number of words, as you would for a magazine.

All told, a program such as RightWriter can help you improve your writing. It includes all the points that earmark good, clear technical writing, and will improve your communication skills. Using this type of interactive program is like having a private tutor and an editor to help you as you write.

SPELLING AND GRAMMAR

Spelling and grammar are important to effective communications, but not nearly as important as getting your message across to your reader. Most editors are not too picky about the spelling of difficult words, nor about the more complicated grammar. Fixing these problems is part of their job.

Spelling or grammatical mistakes can detract from the professionalism of your manuscript. An abundance of these errors obscures your message . . . which is the worst thing that can happen.

Take the time to look up words that you are unsure of or that just do not look right. If the grammar in a sentence appears wrong, have someone else read it and advise you. You can use English reference books if you have to.

With the availability of word processors and spelling checkers, very few misspelled words should filter through to your final draft. You may end up with some typographical errors of the kind I described earlier, so do not let a spelling program take the place of a good proofreading. Software for grammar correction is also available, but not as readily as spelling checkers.

UNITS AND CONSISTENCY

Practically every technical paper will use units of some form; seconds, pounds, volts, bytes, meters, or whatever. To keep your message clear, it is critical that your representation of units be consistent throughout your manuscript.

It is also critical that you use a unit label with every value that requires one. I once had a physics teacher who had a rubber stamp made. Every time you answered a problem with just a number, he would stamp the paper in bold, red ink: USE UNITS. He also subtracted a point from your score, which did wonders for making you remember to put the unit with the value.

If you are writing for a particular magazine and have their author's guide, it will likely have the standard abbreviations used with their publication. If it does not, check a couple of issues to determine what they use.

If you are not sure what magazine you will be targeting, or if you do not find examples of a particular unit, try to use the ANSI standard abbreviation. If you are not sure of the standard abbreviation, use one that you have seen before but be consistent.

The editors will convert any of your units to their standard if yours is different. The magazine maintains an overall style, so if you get a proof back that has different unit abbreviations than yours, do not change them. Even if you think that their abbreviation is "wrong," it is their style, which they will maintain. As with your writing, the most important factor is that they are consistent.

STAY WITHIN THE DESIGN CONSTRAINTS

The design constraints of good English are not difficult to maintain. Although the result is more important than design constraints, you must design within those constraints to have an effective finished product. With writing, your most important factor is the message. If you distract the reader by not designing within the constraints, your message will be difficult or impossible to extract.

Remember that all sentences must have a subject, a verb, and punctuation. Try to use active verbs whenever you can. Use simple words, simple sentences, and simple paragraphs to most effectively communicate your message.

Writing at the proper level for the reader is another essential constraint to ensure that your reader clearly extracts your message from your writing. Using big words just because you know them will not impress anyone and is quite likely to obscure your point. Use the Fog Index or another index to monitor the level of your writing.

Using software to aid your writing can be highly beneficial. Word processing and spelling programs are commonly available and will help polish your finished product. Other writing programs, some employing artificial intelligence, can improve your style and grammar.

Finally, always use units and be consistent. Check the author's guide or back issues of your target magazine to determine what units they prefer.

Stay within the design rules for writing and you will have a superior finished product.

S E C T I O N
THREE

TECHNICAL MANUALS

CHAPTER
6

WRITING AT THE READER'S LEVEL

*With a manual the design problem
is to make the design general enough
that even inexperienced users can
get the full benefit of the manual,
but make it specific and "to the
point" enough that experienced users
will not skip anything essential.*

A good technical manual is not easy to write. Whether it is an instruction manual or a detailed technical service manual, it must be written in such a way that all users, regardless of skill or prior knowledge, can fully use the manual for all purposes that it was intended. Many technical manuals have the same problem as a lot of technical articles—they do not give enough detail.

How do you keep the manual simple enough for nonskilled users, yet at the same time keep the interest of the skilled user so that he or she will read the manual well enough to learn fully to use the product?

With a manual, the design problem is to make the design general enough that even inexperienced users can get the full benefit of the manual, but make it specific and "to the point" enough that experienced users will not skip anything essential. The only successful way to achieve this is to make a "user configurable" manual. By this, I mean that your manual explicitly denotes sections that users of various levels might want to skip and those that everyone should read.

BAD ASSUMPTIONS

Before looking more deeply at solving the dilemma of reaching all levels of readers, we will state some very common, but bad, assumptions made by some manual writers.

The User Will Read the Manual from Cover-to-Cover

Not very many writers *actually* believe this, but how often do you hear a manual writer or a technician say "It's in the manual if they'd just read it." It's an easy excuse to blame someone else for an inadequately prepared manual. Although it is also somewhat invalid, a much better assumption would be "No one ever reads everything in a manual." The way to handle this is to point out very early—in the first few pages—what is important to whom. This is especially true with high tech products. Many times high tech-oriented people read as little in the manual as they have to and learn the rest by experimenting with the product. I do this on occasion myself. It is vitally important that you reach these people quickly and get across the most important information.

There Won't Be Any First-Time Users

Even the most technical product will have first-time users or experienced users who have never used your type of product.

I wrote several manuals for a company that manufactured computer disk drive controllers. These circuit boards were integrated into a computer system through a bus (which is a collection of predefined wires or paths for electrical signals) called Multibus. Since a computer based on Multibus was put together by an OEM* for resale, and, logically the OEM would have already bought a CPU board and memory, the assumption of previous manual writers was that the manual had to tell them how to plug the controller in and how to program it.

What prior manual authors had failed to consider and what lost sales and irate customers subsequently revealed, was that there were two other classes of users, both of which could be

---

* An OEM, original equipment manufacturer, buys components and assembles them into a finished product. Therefore, he supposedly knows what the components do.

considered first-time users. One was an engineer at an OEM who was designing a Multibus computer for the first time. He had a good technical knowledge, but it was either general or specific in other areas, and not specific to busses. He was either doing a literature search or was prototyping a system for the first time.

The second type of first-time user was the end user who had purchased a Multibus computer from an OEM and was adding a disk drive and controller. This user not only did not know about busses, but most likely did not know a great deal about computers other than how to use one.

This first-time user problem was solved in two ways. First, the manuals were rewritten to be user configurable. Second, helpful references were listed . . . right up front where they would do the most good. Two references were generated in-house as application notes. One application note covered bus design, theory, and operation and was aimed at the design engineer. The other application note began with basics of disk drives and storage and, in elementary terms, how a disk controller worked and what its function was within the computer system. A third reference was also listed, a full-length book written by two of the Multibus designers at Intel Corp.*

Thus by configuring the manual differently, as well as providing references, a very complete manual was written that could address practically every user.

Every manual should have some basic information, so labeled, to ensure that all readers begin the material with about the same knowledge level. This could range from a couple of paragraphs to a chapter and possibly references, depending on the complexity of the product and manual.

---

* James B. Johnson and Steve Kassel, *The Multibus Design Guidebook* (New York: McGraw Hill, 1984).

Experienced Users Will Automatically Skip to What They Need from the Manual

Sometimes this is true, but most times it is not. Generally, one of two things happen. One is that the experienced user gets frustrated wading through page after page trying to get to the information he needs *now*. This is where a good index is invaluable. If the writer is fortunate, the user will hang in until he finds what he is looking for. The other possibility is that the experienced user will read just far enough to get the device plugged in and will then toss the manual aside. As the writer, not only do you have to capture the reader, but you have to recognize that different readers have different needs. Your manual must try to solve the needs of every reader you can.

Anyone Knows What the Hardware Looks Like, or Can Find It from the Written Description

Artwork costs extra money in a manual, but customer problems created by the lack of art will cost a company much more in support dollars. Again, try to think of the novice user, because an experienced user *will* know what components look like. And besides, it's not the experienced user that will be calling for support.

The optimum approach would be to picture every item shipped with the product that the manual covers and then to illustrate how each of the parts connect. Actual size drawings are even better when practical; how many people, even mechanical engineers or technicians, can tell the difference between a 6-32 screw and an 8-32 screw without comparing the two? Will all the readers know the difference in appearance between an RJ-11C connector and a BNC connector?

A picture is worth a thousand words (you knew that was coming), but most important, a picture can save your customer service department people a thousand phone calls.

The Readers Have and Will Use the Proper Tool

I have heard this complaint many times from service technicians as well. "The screwheads are stripped! Don't people know to use the right size tool?" NO . . . they don't unless you tell them.

The worst culprits are instructions for power tools, such as lawnmowers. The instructions are almost always kept to a single sheet (apparently it must cost less to fix warranty problems resulting from an underdetailed manual than to print an extra sheet of paper) and the tools list, if any, is often "pliers, screwdriver."

Manuals that have endeared themselves to me have been the ones that have a nice, compact list of materials right up front that you will need to assemble the product. These lists are much more specific and save the person doing the final assembly a lot of time. For many products these days, the final assembly technician is the purchaser (you and me).

The Heath Company (now owned by Zenith) makes electronics kits and has always had a reputation for producing excellent instruction manuals. If you have ever built a Heathkit, you know what I mean. Heath puts a lot of time and money into the manuals, and it shows. Undoubtedly they learned early on that spending the money up front to get a good manual saved much more money in later support. With a kit product, this is even more important because there is no control of what happens in the assembly.

Heath developed a system to ensure that the manuals were not only complete, but were usable by people of all electronic abilities. Once the engineering department finishes a kit, it goes to manual preparation. Then kits and manuals are given to employees from all levels to build (and keep). A given kit might be built by an engineer, a shipping clerk, a marketing manager, and a secretary. This broad range of experience and ability will quickly reveal weak points and inadequacies in the manual. The builders comment on the manual as they work through it, and these can be incorporated into manual improvements.

Heath's tool list even includes illustrations. Figure 1 is an excerpt from a Heathkit manual showing the list of tools required for this particular kit. There is no ambiguity about which

ASSEMBLY NOTES

TOOLS

You will need these tools to assemble your kit.

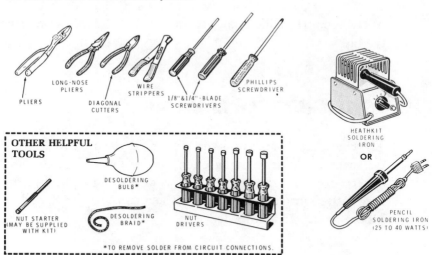

Figure 1

The tool list in a Heathkit manual not only lists the needed tools, but illustrates them as well. Courtesy of the Heath Company.

tools and even which sizes are needed. A list like this should always precede any assembly steps.

I Don't Need to Define These Terms, Every User Knows the Common Ones

This is not quite true either. Most manuals do a good job of defining the somewhat obscure terms, but completely fail to define other, albeit common, jargon. The best place for these definitions, however, is in a glossary. Even experienced users will sometimes not know, or might have forgotten, the meaning of some of the jargon. This is especially true between disciplines. You really cannot expect a civil engineer to know what a 10-MHz 8088 does when you have trouble remembering the difference between cement and concrete.

While a glossary is important to many manuals, it is definitely supportive material. As such, it belongs at the end of the book.

THE CONFIGURABLE MANUAL

The configurable manual is the best way to ensure that all readers, regardless of their background, will be able to get maximum benefit. Actually, the manual itself is not configurable, but the user is given instructions on using the manual that are based on his background and knowledge.

Grab the Reader's Attention Up Front

Put the essential information right up front. It is a good assumption that the reader will begin at the front of the manual.

So you have to make the most out of what you put in the front.

Begin with a good table of contents. An experienced user may jump directly from the contents to the information he feels he needs. MicroPro has found a way to enhance the usefulness of the contents. The manual for MicroPro's popular word processing program WordStar not only has a contents for the entire manual, but goes on to insert a more detailed contents for each chapter. And to make sure the point is picked up by the readers, they point this out on the first page of the chapter entitled "How to Use This Book."

Another technique that MicroPro uses to ensure that the reader sees this material is that they make the "How to Use This Book" chapter a preface. Thus it appears as the first listing on the main table of contents and is the first chapter in the book. It also carries lower-case roman numerals for page numbers to highlight it even more. You can see in Figure 2 how this chapter stands out on the contents. Figure 3 demonstrates how effective the use of a contents for each chapter is.

Leading the Reader through a Configured Manual

To take full advantage of your user configured manual, you have to help the reader through to the sections that are essential for him. Often, this is simply advising the reader whether to read the section on basics (the training, in essence) or to skip to a later section. Microsoft Corporation, a leading software manufacturer, realizes this well. In the manual for their very popular Flight Simulator program, they direct the reader early on page 3:

Because Flight Simulator is a program that will be enjoyed by novices and seasoned pilots alike, this manual begins

1. INTRODUCING WORDSTAR

CONTENTS

Figure 2

The main table of contents from MicroPro's WordStar provides a general description of the chapters in the manual. Courtesy of MicroPro International.

TABLE OF CONTENTS

Figure 3

More detailed table of contents information for WordStar is provided in
individual chapters. Courtesy of MicroPro International.

with an introduction to flight procedures, operations, and control.

If you are new to flying, we recommend that you work through the manual . . . If you have a personal computer, you should read "Aircraft Controls for the PC."

If you are an experienced pilot, you may want to learn how to operate the instruments and controls and then proceed directly to the section entitled "The Editor." You can use the editor to set environmental and flight conditions. You can also use the editor to shift from one of the four Flight Simulator geographic areas to another . . . After you are familiar with the editor, proceed to "Advanced Flight Techniques."

This is an excellent example of how to lead a user through your configurable manual. Note how this excerpt clearly defines what sections an experienced pilot should read, and that the balance can be omitted. Not only will this eliminate some of the burden of reading nonessential material for the experienced reader, but it will also keep his attention, and motivate him to read the sections that *are* important.

Specifications

No matter what your product, a page of specifications up front is essential. If your product is sold on specifications, as many electrical, electronic, and mechanical products are, specifications are essential and belong up front—perhaps inside the front cover.

Quite a few sales are made on products like these via the specifications. Often, potential buyers order a copy of the manual to determine if the product suits their needs. Much of the comparison between products is between specifications.

Therefore, it's not only important to place the specifications up front, but to be sure that you address and present the specifications in the same way as competitors. This sometimes means presenting specifications in a way that does not match your own likes. For example, the gain (ability to increase signal strength) of an antenna is measured in decibels (dB). Because the dB unit is simply a ratio, the gain has to be compared to some reference parameter. Unfortunately, two exist: Gain over a dipole (half-wave antenna) and gain over an isotropic radiator.

A dipole is not only a reference, but is also a very common practical antenna. However, since a dipole has gain in comparison to an isotropic radiator, some manufacturers began using the isotropic radiator as the reference, thereby showing a higher gain figure. The problem is that an isotropic radiator is only possible in theory. Eventually, just about all manufacturers specified their gain versus an isotropic source. While this may not be as practical as a dipole reference, the convergence of all manufacturers to a single reference point makes the specifications of antennas from different manufacturers easier for the user to compare.

Unpacking and Installation

Instructions for unpacking and installing the product should always be toward the front of the manual, following the most important information, such as "How to Use This Book," the specifications, and the table of contents. Once the reader has given you his attention (and it will not be for too long) to communicate the "rules" of the manual, he is going to get anxious to unpack the product and get it going. The next section should be designed to satisfy this curiosity and need of the reader—not to mention that it will be safer if your manual tells him how to do this instead of letting him "wing it."

The unpacking section should not be extensive, and it should not need to be. If unpacking is difficult, the packaging should probably be redesigned, because by the time the user gets the manual unpacked and opened, it is probably too late to give him instructions on unpacking. It is ironic that many products are packed with the manuals on the *bottom*, yet the first chapter is on unpacking! How could the user have unpacked the manual without unpacking the product? This points out the need to coordinate the documentation with the manufacturing department.

Installation, on the other hand, should be as comprehensive as necessary to ensure that the product is installed properly and not damaged. In addition to customer satisfaction, this will save on warranty repairs as well. The biggest problem in the installation section is lack of details.

It is easy for everyone associated with the writing of the manual to assume too much knowledge on the part of the reader. After all, you have probably been working with the product for some time. Some items that come as second nature to you may not be obvious to the reader at all. It is important to take the time to go through the installation process and note each step, no matter how trivial it may seem. Some of the steps will be eliminated, but others will be retained that may not have been included otherwise.

It is a pleasure for me to watch a mechanical engineer put a cabinet together. They always start each screw before tightening the first one, and tighten alternate corners to prevent warping and ensure a good fit. Unfortunately they, as well as most manuals, assume that the user will do this and omit this detail from the manual. Most readers either do not know this or forget to apply the knowledge. I know better, but often forget to install screws this way.

Ehrhorn Technological Operations makes power amplifiers for amateur radio use. The description of installing a cabinet cover on their Alpha 86 amplifier is one of the best I have ever seen:

The top cover must be tilted slightly down at the front when reinstalling it. The front edge of the cover must slide underneath the brushed aluminum top trim piece. Then, the cover is allowed to slide down the chassis and may be slid backward or forward to align the bolt holes. The chassis may warp slightly after the power transformer is installed. For best fit of the cover, do not tighten any of the screws until all are started. Then, tighten screws on alternate sides for the best fit and least warping.

There is no way that they could have been any clearer than that. According to Ray Heaton, ETO's customer service manager, that kind of detail saves them a lot of phone calls which is what motivated them to go into that detail in the first place.

Background Section

A background section should either precede or follow the Unpacking and Installation section. If the user needs a certain minimum set of knowledge in order to install the product, the background section should precede the installation section to avoid problems. The background section needs to begin with a statement telling the reader not only who should read it, but what level of understanding the reader should have when he is finished. You have a goal in mind when you write this section. Do not make the reader guess what it is. The example from Microsoft's Flight Simulator is an excellent way to do this.

As mentioned earlier, references are important to making sure that all readers have the same set of knowledge. The term "Background Section" is really a misnomer. The background portion can be anything from a few paragraphs or pages, to a complete section or chapter, to separate training manuals.

Figure 4 is a good example of a single reference page from Heathkit. You have to be able to solder to assemble an electronic kit. MicroPro's WordStar manual, as is the case with many software manuals, includes an entire training manual with their WordStar products. The user executes a training program on the software diskette to accompany the training. This extensive training availability is an excellent way to make sure that all readers will begin the manual with the same set of knowledge, especially the more advanced sections.

The amount of background material that needs to accompany a manual is not always easy to determine. You can follow the guidelines outlined earlier in terms of defining your user and defining what a first-time user may be. Again, the main purpose of a background section is to be certain that all readers will enter the main section of the manual with the same set of knowledge. This makes the task of writing the balance of the manual easier, because you can now assume a knowledge base and make fewer background references within chapters.

A Quick Start for Experienced Users

Once again, depending on the product, a "getting started quickly" section can be helpful. Experienced users generally want to get the product unpacked and begin using it. If you do not give this user a chance to do this, and with enough detail, he is likely to miss some important features of your product.

I have owned several computer printers over the past few years, so when I bought a new one, I just wanted to plug it in and start using it. Since the manual did not make any provisions for this, I went ahead and did just what I wanted to: I plugged it in and started printing. The problem was *not* that I did not know how to get the printer operating, but that I did not know about, or even that I should look for, the more

1. Push the soldering iron tip against the wire **and** the lug. Heat both the wire and the lug for two or three seconds.

2. Apply solder to the wire and the lug, **not** to the soldering iron. IMPORTANT: Let the heat of the wire and lug melt the solder.

3. As the solder begins to melt, allow it to flow around the connection. Then remove the solder and the iron and let the connection cool.

— Detail 3A —

A GOOD SOLDER CONNECTION

When both the wire and the lug are heated at the same time, the solder will flow onto the wire and the lug evenly. The solder will make a good electrical connection between the wire and the lug.

POOR SOLDER CONNECTIONS

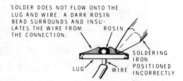

When the wire is not heated sufficiently, the solder will not flow onto the wire as shown above. To correct, reheat the connection and, if necessary, apply a small amount of additional solder to obtain a good connection.

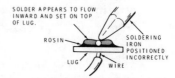

When the lug is not heated sufficiently, the solder will blob on the lug as shown above. To correct, reheat the connection and, if necessary, apply a small amount of additional solder to obtain a good connection.

Figure 4

This soldering tutorial provides a lot of information on a single page. The user is taught how to correctly solder from a single page of drawings and text. Courtesy of the Heath Company.

advanced features. As a result, it was several months before I took the time to go through the entire manual and discovered some very useful features. The danger of this is that I might have bought another printer in this time because I was not aware that the features existed. A quick start section could have let me get printing, but introduced me to some of the more valuable features, plus prodded me to complete the manual.

Some products are easier to use than others, and some have been around for long enough that quite a few people know basically how to use them. Lawnmowers, washing machines, and personal computers to some extent fit in this category.

When you buy game software for your computer, you probably have more motivation to get going quickly than with just about any other product. In the manual for Microsoft's Flight Simulator program, they include a section entitled "Flying Start" designed to let the user fly the program practically as soon as they put the diskette in their computer.

The directions to the user appear right at the top of page 1:

Before you use Flight Simulator, you should be familiar with certain basics. That's what this chapter covers, including a "flying start" procedure for those who want to get off the ground even before reading this manual. To introduce you to Flight Simulator, this chapter describes:

- *The three-dimensional display where you'll be able to view your surroundings from all directions*
- *Flight instruments you use to control your flight*
- *Radios for communicating with the airports where you'll land and take off*
- *Indicators on the control panel to help you monitor your flight.*

Notice how clearly they state what the chapter is about and exactly what you as a user should learn from it. They are also rather clever in doing this as well. The Flight Simulator is not only fun, but it is very realistic. Even as an experienced private pilot, as well as game player, I quickly realized from the Flying Start that I needed to read the remainder of the manual! Such a technique with a lawnmower might not be the best way to handle that product, though. Demonstrating features by dynamically removing anatomical parts would not go over too well with the users!

THE BALANCE OF THE MANUAL

The balance of the manual is mainly the instruction section. Depending on the complexity of the product, the instructions can run from a few pages to a few chapters. This part of the manual might be what you would call more traditional. However, you should still continue to lead the reader through the instructions. There are likely to be many areas that an experienced user can bypass. By showing the experienced user what sections to bypass, you are more likely to keep his attention longer.

Use caution not to oversimplify the instructions section. As in other parts of the manual, it is easy to become so familiar with a product that certain details are skipped over. A good tactic is to let someone else follow your instructions, such as a friend or your spouse, and see if any problems arise. Taking a few minutes to put the polish on a manual will be worth it in the long run.

Theory

Many manuals have a section on the theory of operation. This not only can serve the purpose of providing the manual user with a detailed description of the product and how it works, but two others as well. First, by placing this section at the end (as it would be normally anyway), you have more of a chance of keeping the experienced reader's attention until he gets to this part. Secondly, many of the product's users may well be able to fix small problems once they know how the product works.

You will also most likely want to include such sections as schematic diagrams, block diagrams, wiring diagrams, flow-charts, and troubleshooting charts near the end of the manual. These are always handy reference items and can help to some degree in customer support.

The final section of the manual should contain the service and warranty information. With the state of laws these days, you should check with an attorney and a good book before writing a warranty on a product.

TYING IT TOGETHER

Writing a manual is not what it was a few years ago. More and more complex products have brought more and more complex manuals. In some industries, good—or bad—documentation has made the difference between a product's success and fail-ure. In this market, third-party documentation abounds, further testifying to the quality of the vendor's manuals being poor.

Avoid the bad assumptions outlined at the beginning of the chapter. Explain in detail to the user. Remember that very few, if any, readers will actually read every page of a technical manual.

Make it easy on all users by making a configurable manual.

Structure your manual so that all users will get maximum benefit. Give the neophyte plenty of information to get going and do not forget references in case they are wanted or needed. For the expert, make his busy life easy. Lead him through only those sections that he needs to read to get him, and your product, going.

CHAPTER
7

CONSTRUCTING THE MANUAL

Match quality content with quality production and not only will you have a technical manual that you can be proud of, but you will have a manual that is easy for the reader to use.

*N*ow that you have a format for a configurable manual and have your material gathered together, it is time to construct your manual. You have a great deal of knowledge about the product, so you should be able to write a good manual. However, the problem that often arises when an engineer writes a manual is that because of his or her intimate familiarity with the design and function of the product, details and description of seemingly trivial operations are left out.

The way to avoid this is to solicit input from others who are familiar with the product. This can include other engineers, marketing people, lab technicians, and/or test technicians. This interaction, as with the example in Chapter 6 from the way Heathkit writes its manuals, will result in a superior manual to what could be written by an individual working alone.

Other engineers can provide important information because they have a strong technical understanding of the product, but a slightly different viewpoint than your own. The result is that some possible holes in the heavy technical portion of the manual can be filled in. It is especially helpful to get comments from cross-discipline engineers. For example, an electronics engineer might get some good input from the mechanical engineer that worked on the product.

Marketing people are one of the best sources for input on a manual. In fact, in a lot of companies, it is the technical marketing manager who writes the manual. The marketing manager has a unique perspective because not only is he or she technically competent, but the manager knows a great deal about what the customer needs as well. Quite often, the marketing manager is aware of a customer's need that the engineer might not have thought of.

Lab and test technicians have a valuable perspective for two reasons: First, they have worked with the product nearly as long (and in some cases longer) as the design engineer. Therefore they will have excellent hands-on technical expertise with the product. Second, because their understanding is from a more external viewpoint, their level of intimacy with the product will be less. What the engineer might take for granted, the technician may recognize as important to the end user.

Soliciting Input

Most people are happy to comment on a technical manual because they want it to be easily understood. Try not to impose on anyone who just does not have the time to do it though.

You should talk to the people mentioned above, plus any others that you feel would have valuable information. Get their ideas, and incorporate the best ones into your manual as you write. However, you should treat your first draft of the manual as a prototype. Just as when your prototyping a design, you expect to find areas that need improvement. You might even find an area or two that does not work at all and needs redesign.

Finding these areas is where the people really come in handy. Once you have finished the prototype version of the manual, send copies out to the people that you would like to comment and critique the book. There is an art to this, though. When you give the people a copy of a manual, tell them sincerely to make any comments they want. If you do not allow freedom for their input, you will not get as high a quality information. Most importantly, establish a deadline for their comments and make sure they understand it. You can still expect some to come in late, but if you do not establish a deadline, you will not get some back at all.

Once you get the comments back, go through them carefully and with a thick skin. Look at them as objectively as you can; there will be some good ideas suggested. You do not have to use all of the comments of course. Once you have incorporated the comments into a second draft, it is usually a good idea to circulate the document for comments. This time, make the deadline much shorter. You are looking for "touch-up" comments now.

When using this technique, be sure and set a limit to the number of times that you will send out the documents for comment. The two above are usually sufficient. If you do not do this, you can get into a never-ending circle of comments and revision.

I once wrote an application note for a company I worked for, and it was distributed to three other people, including my boss and the vice president. The draft copies came back with comments from everyone, with quite a few from the vice president. I incorporated the comments and redistributed the papers. Again they came back with an abundance of comments, which I incorporated. I then gave the papers to the Marketing Communications Manager, and she distributed them again! Once again they came back with comments. This went on for three more rounds. I finally threatened to quit if they were distributed again! The point here is twofold. First, you have to decide when you are finished and stop. And second, every time you distribute a paper or manual for comments, you will get them . . . it is guaranteed. *You* have to be the one to control the comments and shut them off. By stating up front that you will allow two go-rounds, you make it explicit when you will deem the project finished.

When you have your final draft complete, it is important to have one more read. However, this last read should be strictly a proofread and should not change content. It is best to have someone proofread your manual who is not familiar with the manual or the product, concentrating entirely on the grammar and spelling.

ORGANIZING THE MANUAL

Much of the last chapter was devoted to organizing the manual in terms of what material to put in the front, making the manual user configurable, and so on. There is a bit more to organization that occurs during the construction of the manual. First, you need to break the material down into logical sections.

The beginning of the manual has already been defined. The "meat" of the manual is the actual set of instructions that tells the user how to use the product or describes how it works. You need to decide the best order for the presentation of this information and then divide the material up into logical chunks.

You can begin breaking down your material in the same way that you wrote an outline. In fact, you should be able to use your outline as a good starting point. Your first attempt might be operation and maintenance. Next, if you are writing a manual for, say, a computer, you might break operation down into hardware and software. Then, break down each of these two subsections into the various functions that they perform.

If your product is not too complex, you will probably want to describe each section in its entirety. For more complex products, it is best to describe a particularly difficult function in overview, then later describe it in full detail. This will give the reader a chance to learn a little about the function before you get too complicated. It also enables experienced users to skip the elementary section and go straight to the more detailed section.

Number the Pages and Sections

Once you have the material divided into the sections that you feel comfortable with, you should number the chapters, just like you have on your outline. Some people prefer to number each subheading, so that a typical table of contents might begin as shown in Figure 1.

I find this is an extremely cumbersome way to write a manual. It is done for the writer's convenience, rather than the reader's (although some military manuals benefit from this style). There may be several numbered subheadings on a single page and wading through them can be confusing and can take the reader's attention away from the material. Very few good manuals number beyond the chapter itself, just as this book is numbered.

Internal numbering used to be a requirement of Army and Navy technical manuals—the Air Force numbered paragraphs (1-1, 1-2, etc.) in each chapter. With computer-aided indexing, perhaps the DOD has abandoned subhead numbering. While this may have been useful for sifting through the tons of documents required by the government, it is a cumbersome way to have to read a manual about a consumer product.

1.0 Introduction
1.1 Unpacking
1.1.1 Removing the product from the box
1.1.2 Removing the packing screws
1.1.3 Unpacking the power cord
1.1.4 Inspecting for damage
1.1.4.1 What to do in case of damage
1.1.4.2 Save the original carton
1.2 Initial checkout
1.2.1 Power on check
1.2.1.1 If the unit doesn't power up

Figure 1
Some manuals use numbering *ad nauseum*, which does more to confuse the reader than help him. This style was adapted from military manuals, and is not appropriate for technical manuals for consumer products.

The reason this is a convenience for the writer is that many word processing systems can automatically index using these subheadings. It is a less than optimal index, because the reader must look for the subhead number rather than the much easier to find page number. This technique is, thankfully, being used less and less.

An easier way to help the reader to follow is to number the chapters, as suggested above, and use subheadings, exactly as was done with the technical article. The reader is more accustomed to reading material laid out this way, so it will not require him to adjust to a new style, which is another distraction factor for his attention.

Use a two-part page numbering system within the chapters. Begin with the chapter number, a dash, then the page number, such as 2-33 (Chapter 2, page 33), or 4-9 (for Chapter 4, page 9). Doing so will enable the reader to locate chapters easier. Plus, if you have to revise pages later, it will be easier to incorporate them into the manual since you will only be changing the page numbers within the single chapter.

Supplemental Material

The supplemental material goes in specific places, although you still have room for creativity. The table of contents is the first item in the manual. Use a detailed table of contents, or alternately, use a less detailed table of contents to start and a more specific table of contents for each chapter. The WordStar manual had an example of this technique in the previous chapter. Another WordStar technique detailed in Chapter 6 was the use of a preface to grab the reader's attention.

The preface is the next section in the manual, and directly follows the table of contents. Because it is the first piece of information following the contents, it is listed first in the contents. The pages of the preface are also numbered with lower-case roman numerals, which also draws attention to it. You can use these creatively and to your advantage by putting your most important information here, and letting the reader know it in the table of contents.

Up to three sections may be included at the end of the manual. These are the glossary, appendix, and index. The glossary isn't a requirement for a technical manual, but it is a handy feature. Neophytes can find words that they may not be familiar with, and more experienced users can find that word that they just cannot remember.

Appendices are important to most manuals. On occasion, you may want to put the specifications in an appendix. Reference tables, drawings, and schematics usually belong in the appendix as well. Sometimes a troubleshooting section is included in the appendix with appropriate service telephone numbers.

The index is one of the most important components of the entire technical manual. A good index is referred to just about every time a user picks up the manual after his first reading. A section on indexing follows later in the chapter.

FIGURES

Figures are extremely important in a technical manual. Research done on speeches shows that over half of the message in a speech is conveyed in the visual aids used with the speech. Figures play a big part in a technical manual as well.

Because figures play such a big part in a manual, it is worth having them done professionally. If you are using drawings, have the drafting department in your company draw what they can, and have a professional artist draw what the drafting department cannot. High quality drawings make understanding much easier for the reader and add to the credibility of your product.

If you are using photographs, spend the time and money to have a professional photographer do the work. Nothing cheapens a manual more than to see poorly done photographs. For the type of quality that you need in a manual, 35 mm is almost never suitable. 6 × 4.5 cm is about the smallest format that will yield quality results. 6 × 6 cm or 4 × 5 in. negatives will give the superb quality that you see in magazines and other high quality manuals.

To save cost with a photographer, decide what you will want photographed and design a plan to use the photographer's time most efficiently. Be sure to take all the equipment that the photographer will need. Nothing runs up the cost of a photo session more than having to start the whole thing over again. It is a good idea to take back up gear for anything that could possibly malfunction. Remember extra light bulbs, fuses, and batteries.

Describe each scene to the photographer, but listen to his suggestions. Photographers can often make suggestions that will save you time and money, and may make your photographic message more clear as well.

With both photographs and drawings, it is important not present too much detail. Write down a design goal for each illustration, and try to make the illustration achieve that goal with as little unnecessary detail as you can. Anything that does not directly lead to achieving your goal should be omitted, as it distracts from your principle goal. If you can afford it, the advice of a graphics designer can help here.

While you should not include unnecessary detail in the illustrations you use, you should use illustrations freely throughout the manual. It is always easier to describe a specific part or function with an illustration than with words. Combining the two will give you the best chance for the reader understanding your instruction.

Captions and Callouts for the Figures

Each figure should have a caption that describes it fully. The captions should be in complete sentences and provide the reader enough detail so that he can understand the figure reasonably well from the caption alone.

In addition to the captions, figures in technical manuals need *callouts*. Callouts are labels of important parts that appear within the figure. You should provide enough callouts for the reader to easily identify the assembly, individual parts, or whatever is necessary to make the figure useful. All parts that are referred to in the text should have a callout on at least one figure. A good example of a figure with callouts is Figure 2.

Refer to the Figures

You should always refer to the figures within the text of the manual. This accomplishes two purposes. First, you will lead the reader to refer to the figure at the appropriate time with the text. This will give you the maximum benefit from the figure. Second, many readers will use the figures alone if they are reasonably familiar with the type of product. If they need more information than they extracted from the caption of the illustration, the figure reference in the text will enable them to find a more detailed description.

Use the callouts on the figure in the text as well. For example, using Figure 2, the text could read:

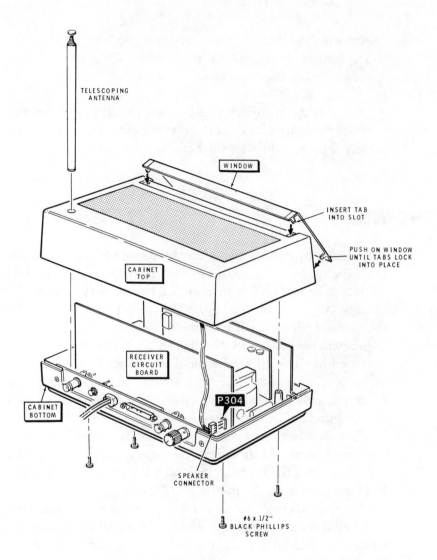

TELESCOPING
ANTENNA

WINDOW

INSERT TAB
INTO SLOT

PUSH ON WINDOW
UNTIL TABS LOCK
INTO PLACE

CABINET
TOP

RECEIVER
CIRCUIT
BOARD

P304

CABINET
BOTTOM

SPEAKER
CONNECTOR

#6 x 1/2"
BLACK PHILLIPS
SCREW

Figure 2

Callouts are the labels for important features of a figure. This is an excellent example of using callouts to label the key parts in a drawing. Notice how clearly all the essential parts are labeled. Courtesy of the Heath Company.

Assemble the cabinet using the four #6 phillips screws. Then insert the telescoping antenna through the top of the cabinet and screw into place.

A figure index is often a help to some users as well, because they use the figures without the text at times. The figure index usually appears right after the table of contents.

THE INDEX

The index may be the most important section in many manuals. Once the reader has been through the manual once, most of the references he makes from then on will begin with the index. For example, suppose you use your VCR as the tuner for your TV set, as many people do. If you move to a new city, you will want to program new channels into the VCR. Unless you have programmed in channels recently and remember how, you would undoubtedly get your VCR instruction manual and go to the index and look up channel programming. You would be quite disappointed if it was not listed.

Troubleshooting is also best accomplished with the help of a good index. Most indexes will have common problems indexed so that the user can look up the symptom in the index and turn directly to the page.

You do not want to disappoint your user any more than you would want to be disappointed by your VCR manual. So, take the time to do as thorough and useful an index as you can. You can ask for others to help index as well. They will often index useful items that you may not have thought of.

Your index should be the converse of your figures: Go into as much detail as possible with the index. It is just about impossible to make an index too detailed. There are several techniques to indexing that will improve the usability.

Cross-reference items within the index. Your VCR manual might have a cross-reference under Programming that reads "*See* channel programming." It is always a warm feeling when whatever phrase you choose to look for in the index is there. Take time to list as many cross-references as you think are necessary.

Be sure to avoid circular references, however. In the example above, the entry Channel programming should NOT say "*See* programming!" This can happen by accident or change of plans and is absolutely infuriating to the reader (as well as useless).

A sub-index is also a valuable tool. A sub-index is used similarly to the way a subject card is used in the card catalog. Your VCR manual might have a sub-index under Programming which read:

Programming:
  Channels, 24
  Clock, 53
  Timer, 55

Using sub-indexes will save your reader time and is another way to make your manual more useful, and its use more efficient.

Use boldface and italics to help the reader. It is best to use boldface with sub-indexes. This will draw the reader's attention to the subject heading and allow him to get to the topic faster. This becomes more important as the index becomes longer and more complex. While a comprehensive index is desirable, you have to help the reader get through the more complex index or it will lose its value. Italics are used for a portion of the cross references, such as:

Channel programming; see Programming

Many word processing programs have indexing capability. Take advantage of the feature whenever you can. However, remember that if you are having your manual typeset, the page numbers will change, so you will have to doublecheck each one to get the correct page number once the manual is in galley form.

Another computer technique is to use a database program. You can simply enter a topic and page number each time you come across one. Then, after you have finished the entire manual, sort the topics and combine duplicate references by listing all page numbers under a single listing. For example,

Programming, 22

Programming, 43

Programming, 67

would become:

Programming, 22, 43, 67.

Proof the Index Carefully

I am sure just about everyone has had an experience where they looked up an entry in the index only to find that the reference to the topic did not occur anywhere near the page that it was supposed to. This is quite frustrating for the reader, and is one reason to proofread the index carefully. Since an index is rather boring to proof, the result can be that many errors are made or overlooked when proofing.

The best way to proofread an index (or any document for that matter) is by two people. One person reads the index, while the other looks up the reference. You can use an index person and a person for each chapter to really speed things up, and also to limit the errors somewhat since each person will not get as tired.

PRINTING

The method that you use to print your manual depends on your budget and what you want the final product to look like. Depending on your budget, and the length of the manual, you may opt for printing directly from the word processor, using a desktop publishing system, or having a professional printer typeset and print.

Small manuals can be printed directly from a word processor without too much trouble. Integrating figures into word processor output is not too easy. Also, the single column format of the output is more difficult for the reader to follow than multiple columns that can be formatted by a desktop publishing system or by a publishing house. For optimum results, do not print long manuals (over 5–10 pages) by photo-offset printing directly from word processor output.

Desktop publishing systems allow a lot of customization to be done to the output of your document. You can integrate figures, change type fonts and sizes, and divide the page into multiple columns. Multiple columns allow the reader to read a shorter line of text before moving to the next line, which is easier to do. You have probably noticed that you more often make an error when returning to the beginning of longer lines than shorter lines. This is one reason why magazines and newspapers use multiple columns per page. Books use smaller pages for the same reason.

You should take advantage of the desktop publishing

system's column formatting. The number of columns you use depends on the size of type. If you do not have a laser printer, two columns is probably about the best compromise. If you have a laser printer, three, or perhaps even four columns will work well. Experiment with different numbers of columns and find out what works best for your format.

A laser printer and a desktop publishing system also allow you to easily integrate figures into a manual. You can make many of the figures directly with the desktop publishing system, and insert them right into the output as it is printed. You can also print directly onto both sides of the page.

The most professional looking manual will be produced by having the text and figures printed by a professional printer. Most printers these days can accept printed hardcopy, as they always have, but often will give you a discount on typesetting if you submit the data on a floppy disk. This is because they do not have to enter the copy themselves, which is one of the most time-consuming processes of the printing process. Check with printers in your area to see if they discount, and what formats and word processing programs their systems are compatible with.

You will first receive a galley proof from the printer. A galley is the first printing, and usually has only one or two columns on a page. It is especially important to proofread the galleys carefully because any error that you do not catch will end up in final print. Most typos that you see in magazines would have been eliminated by more careful proofreading. Have two people proof the galley by reading together. One person should read the original manuscript aloud while the other reads against the galley. This not only improves the efficiency of the proofreaders, but is the only way to assure that no letters, words, lines, or even entire paragraphs were left out. This occurs from time-to-time and is quite embarrassing if it is not caught.

Once the galleys have been proofed and approved, the printer will print the final copies. These will look much like the galleys, except they will be on a film type material. At this point, you should employ a professional paste-up artist to lay out the manual. Large printshops probably have a paste-up artist on staff and can provide the service for you. A paste-up artist can also provide you with justified (lined-up) tops and bottoms for your pages, the final touch of professionalism.

Clearly, going through a printer and paste-up artist is not inexpensive. However, the results are by far the most professional you can attain. If you want a professional looking manual, and have the budget, then a professional printer is the way to go.

PACKAGING THE MANUAL

The packaging of your manual depends on the same criteria as the printing. If your manual is small and was word processed, you might just want to staple it together. If it was done on a desktop publishing system, you might want to put it in a three-ring binder. A printed manual would look best if it was bound.

Stapling is alright for a short, word processed manual, but stapling almost always results in a difficult-to-read upper-left corner. Even a short manual can be packaged in a way other than stapling.

Using a three-ring binder has quite a few advantages. For one, you don't have areas that are unreadable, providing you printed with enough margin to allow three-hole punching. Another thing is that the binder provides protection for the fragile paper content. This becomes more important as the manual grows larger, and its weight increases.

A three-ring binder also allows for updates to the manual to be easily added. For a product that has been around for years, this isn't too important. For many products, especially computer hardware and software, updates are fairly common. A three-ring binder allows the user to easily add the updates that you send him.

The most professional way to package a manual is to have it bound. Spiral binding is the simplest way to bind a document, and several inexpensive machines are available to enable you to use this technique on output from a desktop publishing system.

A printer can bind your manual into a very professional looking end-product with glue. The pages are glued at the spine, and a semi-rigid cover is then glued onto the spine of the pages. This is the best looking manual and provides good protection as well. For large quantities, this is also the most economical method of packaging because it is the least labor intensive process.

CONSTRUCTING THE FINAL PRODUCT

Constructing the final manual requires careful planning as well as input from other people. Use the resources you have available to comment, make suggestions, and proofread. However, don't let the project get out of hand: Establish a deadline for each portion and stick to it.

Break the material down into logical, digestible sections. Use numbering for the chapters, and make good use of sub-headings. However, don't overnumber, as it can distract your reader's attention away from the message that you are trying to communicate.

Do not skimp on the number of figures. Combining good illustrations with good text is an unbeatable combination. Use professional artists and photographers for the best results.

The index may be the most important section of the entire

manual. Take the time to do a useful, detailed index. Use cross-references and sub-indexes to facilitate the reader's use of the index. Remember that most references to the manual after the initial reading will begin with the index: Do not let the reader down with a poor one.

The printing and packaging are the last two areas of concern when constructing the manual. Match the sophistication of the printing and packaging methods with the size of the manual. Small manuals can be printed directly from a word processor and stapled together. Larger manuals can be printed using a desktop publishing system and either placed in a three-ring binder or bound using other techniques. For the most professional looking manual, use a professional printer and paste-up artist, and have the manual bound. Glue-type binding is the least expensive way of producing a large quantity of manuals.

Match quality content with quality production and not only will you have a technical manual that you can be proud of, but you will have a manual that is easy for the reader to use.

S E C T I O N
FOUR

TECHNICAL SPEAKING

CHAPTER

8

> ## *PREPARING A*
> ## *TECHNICAL*
> ## *SPEECH*

*The key to giving effective technical
speeches is to prepare well for them,
which entails research, questioning,
and being open and honest.*

*T*here are certainly fewer people who enjoy giving speeches than there are who enjoy writing. Just as you will inevitably have to do some writing (even if it's just a now-and-again memo), you will also have to do some public speaking, even if that is only giving a progress report to two or three co-workers, or asking your boss for a raise.

This chapter deals with technical speaking, so you will have to refer elsewhere if you fancy yourself as an after-dinner speaker as well. During a technical career, be it engineering, marketing, or management, you will probably run into one or more of seven different types of technical speaking—not necessarily speeches.

Presenting Technical Papers

This is what most people think of when the term technical speech is mentioned. The typical technical paper presentation takes place at a seminar, forum, or tradeshow, and often consists of a group of engineers being bored stiff by someone reading a paper that the speaker authored on a relevant topic. Of course, you are bored because you have a copy of the paper on your lap, and probably got tired of people reading to you years ago. There are better ways of presenting technical papers. The better ones are the ones we will discuss and attempt to emulate.

Panel Discussions

Closely related to a technical paper presentation in format is a panel discussion. Its format is similar in that they occur at the same type of seminars, forums, and tradeshows. What you prepare, the way you prepare that information, and the way you present the information are very different from preparing a technical paper.

Design Reviews

A design review can be a horrible nightmare or an intelligent exchange of facts. Although the personalities of your superiors have more than a small role in which outcome you experience, you can place the odds in your favor by properly preparing.

Board Meetings

Personally, these types of technical speaking have always been the most frightening. Although a design review can be a bad experience, at least you know what the questions probably will be. In a board meeting, you can prepare to describe why the goniometer is two weeks late, only to be hit with "Why did we have to spend 25 percent more on drafting last month?" Careful analysis of other's motives are needed for this type of speaking.

Product Demonstration

I have always enjoyed product demonstration speeches. They are like show-and-tell from grade school, where you get to bring in your new toy and tell the rest of the class how wonderful it is. The key ingredient here, of course, is *believing* how wonderful it is yourself.

Sales Talks

While this book is not written primarily for salesmen, most technical people either serve as a pseudo-salesman on occasion, or make calls with salesmen from time-to-time as a technical expert. The biggest problem with technical people in the field is that they talk too much, tell too much about the product, and on occasion "unsell" a client who has already been sold. This is according to the experts—the salesmen who are in the field every day, and have to retract later your promises that can't be met.

PRESENTING A TECHNICAL SPEECH

You may be asking the same question about presenting a technical speech as I assumed you would ask about a technical article back in Chapter 3: Why would I want to subject myself to this torture? Basically, the reasons are the same: To gain positive exposure for your company, its product, and yourself. There is another very good reason for presenting a paper at a conference, it provides a benefit that you will not get from a published article. You will get to personally meet many people with similar interests, as well as experts in your field. And if you have prepared well and made a good presentation, you will also now be recognized as an expert.

Some of the best contacts that I have made have been at shows and conferences, and many were as a result of a technical presentation that I had made. Often these contacts will do you a lot of good later—or sooner—in your career.

There is another good reason for presenting a technical paper, but it is generally a taboo subject. That is, it looks very good on your resume and comes in handy at review time. It shows that you can and *will* make the extra effort to further your company as well as yourself. It is odd that this benefit is not usually mentioned, since it often produces opportunities and puts money in your pocket.

You should approach the technical presentation exactly as you did the technical article. If you need a refresher, or skipped directly to this chapter, review Chapters 1 and 2.

Begin preparing for your speech with your specification—the outline. When you begin your outline and research, forget about the length of the speech. If you concentrate too heavily on staying within the time constraints, you probably will not cover your topic as effectively as if you get the information first, then organize it to fit the time allotted.

There are two formats for technical presentations at conferences. In the first, you write a paper and submit it long beforehand. If your paper is selected, you have the opportunity to present it at the conference, as well as having the paper published in the *Proceedings* book that is distributed at the conference. The second format is one in which a technical speech is given, but no formal written copy is distributed. The way that you prepare for these is somewhat different.

For both methods, for that matter for any speech, using 3 × 5 or 4 × 6 notecards is essential. But the way that you use them is important. Notecards are for notes and reminders. Like an outline, they keep you on track, make sure you cover what you intended, and guide you as you progress through your speech.

Speakers who read from notecards are quite dull. Because they are reading instead of speaking, they seldom use the inflection, emphasis, and gestures that make speaking interesting to listen to. I make it a point to never write a complete sentence on a notecard. That way, I can't read from the card even if I try.

Notecards are simply reminders and guideposts. If you have had time to prepare well, you probably will only glance at the cards a few times. If, like most of us, you have had to put together the speech with the help of several pots of coffee, they can be referred to more frequently, but still as reminders.

Notecards can also keep you from getting lost in the middle of a speech. This is incredibly embarrassing, but can easily happen. Several times I have been so intent on making a point that when I finished, I had no idea of what I wanted to talk about next. A quick glance at my notecard and I was off with an apparent pause for emphasis. That brings up an important point: It is important to leaf through your notecards as you speak, even if you are not using them. Great notes that are under four cards when you need them are not of much value.

Presenting a Published Paper

If you are preparing for a conference for which you will have to submit a written paper, you need to follow the advice of Chapters 1 through 5, and write the paper as you would any article. This is important because your paper is not your speech, and your speech is not your paper. You should write your paper first, and write it as if you were not going to present it.

Many times, if not most, papers that appear in *Proceedings* are practically useless once the attendees get home. If the paper was written from the speech, there will be so little detail in the published paper that it will be impossible to make use of it for real research and learning. The paper should be written so that if the *Proceedings* is read a month or a year later, the reader can extract the knowledge from the paper that you tried to pass along in your speech.

On the other hand, the speech portion of the presentation must be written as a speech. If you write the paper first, it makes an excellent set of notes, but that's it. You should never read your paper; it is generally too long for your allotted time, and it is always too boring. The reason you were asked to give the speech is because a live body is more interesting than a written piece of paper—so add some life to your topic.

Thus your paper and your speech must be closely interrelated, but distinctly different. If you want to race on Saturday, and haul wood on Sunday, you better buy a car and a truck because a compromise will be disappointing. In this case, your compromise will disappoint either the conference attendees or the future reader, depending on which way you have compromised.

Once you have finished your paper (complete with illustrations, captions, and references to the illustrations), you are ready to begin your speech. Most speeches need visuals too. A visual aids your presentation, and should not be used to take attention away from yourself so you do not feel self-conscious. There will be more on visuals in Chapter 9.

The best way to write your speech from your paper is to make notes on cards as you read back through the paper, exactly as if you were using your written paper for researching. Your notes will then follow the same chronology as the paper, but when you deliver the speech you will use speaking terms instead of writing terms, and that is the whole key to this type of a presentation.

Let me give an example of this to make it clearer. Here is a sample paragraph from a paper that might have been written for a conference on computing.

Microprocessors were evaluated during the early research phases to select the optimum processor for the application. The primary criteria were total parts count, speed of execution of a benchmark written to emulate the application environment, and the ability to run a high level language (HLL) such as C. The Motorola 68010 and the Intel 80386 were evaluated within these constraints.

You could read this verbatim to your audience, and it would not be too bad, but I am assuming you bought this book and are taking the time to read it to be better than "not too bad." Realistically, you would never speak that way if you were telling an old college buddy how you accomplished this marvelous feat. Your notecards from this paragraph might read:

- evaluated micros first
- needed low parts count
- used own benchmark
- needed C support
- eval. 68010 and 80386

From these notes, your speech might sound something like this:

> *Our first task was to select the microprocessor for the product. Because of size constraints we wanted to have a minimal parts count, but we didn't want to sacrifice performance just for an IC or two. So we wrote a benchmark that would exercise the processors in a way that would emulate their actual use. No canned benchmarks—I don't think they give accurate data for a specific application. And since our programming staff likes to program in C, and they write very efficient C programs, we wanted the processor to be efficient in C as well! We ran Motorola's 68010 and Intel's '386 through these benchmarks, and the results are on the graph behind me.*

I think you will agree that using the speaking tone, and speaking from notes is much more interesting from a *listener's* viewpoint (as opposed to a *reader's* viewpoint). The important changes when progressing from the written material to your speech are: (1) use less formal terms (Intel's '386 rather than Intel's 80386 microprocessor), *talk* to the audience (the graph behind me), and (2) use short sentences . . . even sentence fragments and phrases because that is exactly the way we talk. It is also important to use hand gestures, as you would talking to a friend, and even move around a bit if that is more comfortable. Because you will be working from notes, these conversational nuances will all be much easier than if you were reading from a page.

One quick point: Humor almost never works in a technical speech unless it is spontaneous, so do not spend much effort preparing a lot of humor. If something strikes you as humorous as you are giving your speech, toss it in, but move on quickly if it does not work.

A friend of mine was asked a question following a speech about the high price of IC chips. Borrowing from a line from a popular song by Dire Straits ("Money for nothing, and your chicks for free"), he concluded his answer with "money for nothing, and your *chips* for free." It was spontaneous, and it worked.

A Verbal-Only Presentation

Giving a verbal speech without submitting a paper is not as easy as it first appears. You still have to specify, research, and refine your ideas until they are in a presentable form. While you do not have to actually write the paper, there is an advantage to having to write one: Each time you read what you have written, you are in effect giving the speech in your head. This practice really does help you present the speech.

You should still start with an outline to serve as the specification for the speech. The speech will easily go together from the outline.

From your outline you can use note cards to write down important ideas and concepts. You can also add cards from any research that you may need to do. Once you have completed a stack of cards that adequately represents the relevant information about your topic, you can design your speech using the cards as components.

Using the cards from each topic separately, you can easily rearrange the order for a better flow, and omit topics that don't fit in the flow of your speech. Then you can combine all of the topics into your draft speech.

Using this draft as a prototype, go through card-by-card and in your head, give the speech to yourself. The first pass through is usually a little rough around the edges. Ideas will come to you as you think about what to say about each card. Write these ideas down so you will be certain to make that an important point.

When you have gone through the speech a few times, you should have a set of notes good enough to repeat your finished product in front of the audience. You will also want to make notes or fill out separate note cards with ideas for visuals as you think of them. You will want to use a separate slide for at least every three or four cards, and sometimes a different visual for every card.

GETTING THE RIGHT LENGTH

You will almost always have a time limit for presenting your technical speech. If not, you should set one for yourself. Twenty to 30 minutes is about the standard time for a technical presentation.

Once you have your speech reasonably complete you should go through and determine two things before timing the speech: (1) Do I focus on my point and cover it well? and (2) Do I try to cover too much and not really make any point? The first should be answered "Yes," and the second "No." If you can't solidly answer this way, you should do more research and refining before you see if it fits. A 30-minute speech that never makes a point is useless.

When you are satisfied that you have good raw material and stick to and make your point, then you should estimate its length. A good rule of thumb is to allow two minutes per note-card. This is only an estimate, but it is a good place to start. If you have 50 cards for a 30 minute speech, you probably have too much information, and if you only have 5, your speech is probably going to be too short.

To get more exact timing, look at each card and the visual that you have planned. Will you talk two minutes about this item, will it be more or less? Go through what you had constructed in your head and time it—or even better, speak the section aloud and tape record it. You will usually speak about 30 to 50 percent slower than you will "think" through a speech, so if you do not taperecord yourself scale your timing appropriately. When you have a more accurate estimate, note it and then repeat it for each card.

Then, select cards to equal roughly the time allotted. You may have to leave some topics out of your speech to meet time constraints. If so, analyze what you have and eliminate cards that either do not add to your point, or their absence does not detract from your point.

The final step is to practice the speech and time it. This is not exact, because you will present differently in practice than when you are on the podium. When you get close to the allotted time, you are done! There are more details on presenting, and staying on time in the next chapter.

WHAT HAPPENS WHEN YOU RUN OUT OF TIME?

It is tough to run out of time preparing for a speech that you have already submitted a paper for, but of course, it happens. You are much more likely to run into trouble because of procrastination for a speech that you are giving verbally (only). You can also run up against the deadline when writing the paper for submission as well. What do you do then to still come out with good quality (you will not be able to get excellent quality without some work)?

Late with Your Paper

When you are getting close to your deadline (or even if it is already passed), an outline is even more important than it is normally. You can write more quickly with an outline, because the organizing you did with the outline will save you time wasted reorganizing. If you just write from your head trying to save time, you will most likely have to do some editing because ideas seldom come to you in the most logical flow.

However, if you work from an outline, you can write the paper by expanding the outline in full sentences on paper. You will still need to polish things up some, but the ideas will be organized in a clear path and you will have written a better paper. As with a computer program, it is *much* easier to correct syntax errors than it is to fix an error in logic.

If you run into a problem of being seriously late, submit your paper without the figures if they are not done. By all means call the editor or program chairman and let him know your intentions. Also commit to a date and deliver your figures on-time. Most likely they will need time with your text before they actually need figures, and getting your late text to them as soon as possible with help.

Late with Your Speech

This is all too common. If you are running late, by all means work on the visuals first. Whatever type of visual you are using, it will take longer to make than the rest of your speech will require. Begin with an outline, and take your figures directly from it.

Once work has begun on the visuals, you can use those to construct your notes. Simply write down, on note cards, the key points that you will want to say about each figure. If a point that requires more research is not critical, cover it in a different way, or you can bring out the point and note that it needs more research. If you need to do research, do it as soon as you have a handle on what needs more information. Plan your research and do it quickly.

Leave yourself enough time to go through the speech in your head to make sure it flows well. Double check your note-cards and your visuals. When you are in a hurry, you are most likely to make a mistake that will embarrass you later.

THE PANEL DISCUSSION

You do not have to be an expert to participate on a panel. You do not even have to know a great deal about the subject. In fact, you do not have to know anything about the subject . . . or so it seems at times when a person sits on a panel but never speaks. Seriously, you do not have to be an expert, but you should prepare well enough to offer pertinent comments when the opportunity arises.

Your design goal when participating in a panel discussion is to establish yourself as an expert on the subject. They key to achieving this goal is in the preparation.

Preparing for a panel is different from preparing for a speech because you cannot focus as sharply on the topic. In electronics, it is similar to a black box problem, where you are given a black box enclosing a circuit. Through application of theory and judicious use of test equipment, you have to determine the circuit inside. Surprisingly, it is not as difficult as it seems.

Many panel discussions use seed questions that a moderator will use to begin the discussion or keep it going if questions from the floor get slow. Use the seed questions to your advantage by preparing enough to offer a comment on each one.

To prepare for a panel discussion, use your available "test equipment." Start with yourself; what questions would you ask if you were in the audience? What would you want to know about your competitors? They are likely to ask the same questions about your product or company. If you are well prepared, you will come across very favorably to the audience—a position you want.

What are the subtle problems that exist in your company or product, or in the technology? Be brutally honest, because if you are not prepared and a question is asked about the subject, you and yor company can lose credibility.

When you have finished asking questions, pick up some recent trade journals. If your panel will focus on your product, what does the competition's advertising focus on? Do they tout a feature that you don't have? If so, design a plan to counter their arguments. Does their advertising avoid anything that you will want to mention?

Whether your panel will discuss products or technology, read a few recent articles and editorials on the subject. What do authors and editors perceive as problems and strengths? Address the problems and capitalize on the strengths. Even if you don't agree with an editor's perception of a weakness, many of his other readers will, and this will be your opportunity to bring them over to your side.

If you do not check out your competitors' advertising, nor look up to see what the press has been saying, you will at best be missing an opportunity to win support. At worst, you could make yourself and your company look out of touch.

The final step is to talk with colleagues and let them ask questions. Talk to the engineers who helped design the product or who work with the technology. Talk with people at other companies who use the technology or your product.

By all means, if your discussion will deal with your product at all, talk to one or more of your company's salespeople. The salespeople are in contact with the users of your product every day. They will usually be anxious to tell you what customers want to know about the product or what they perceive as its strong points, and what they perceive as its weak points.

By this point you will have a body of knowledge to make you an expert—at least for the sake of the panel. Even though you will not be able to comment on every topic brought up, you will be aware of most of the subjects and should be able to offer quality comments on them.

Pierre Salinger, Press Secretary for President John F. Kennedy, stated in a recent interview that this method was exactly how he and Kennedy would prepare for a press conference. Salinger would compile about 100 questions that he felt might likely be asked. Then the two would go over the possible questions prior to the conference. Salinger, an experienced journalist, had the same advantage in compiling likely questions that you do—he had participated from the other-side-of-the-fence.

Fielding Questions

It is not enough just to be well prepared. The way you field questions and the answers you give have an influence on how well you promote yourself, your company, and your product.

If you have a good answer to a question, field it as soon as it is asked. Speak clearly, confidently, and at a fairly rapid pace. Recent research has shown that speaking rapidly, but not too quickly, transmits a message of confidence. By fielding the question quickly, and with a good, confident answer, you have made your answer the benchmark for other panelists' replies—and this put you in an offensive, rather than defensive position. As the great Green Bay Packers' coach Vince Lombardi put it, "The best defense is a good offense."

The other panelists, who are likely to represent several differing viewpoints, will have to *add to* what you have started, thus tacitly making you the expert. Should someone else make a good challenge, as they likely will, be prepared to rebut quickly to keep your position strong.

If you do not have a strong point to make about a question or topic, allow someone else to take the lead. If they lead with a strong point, look for an opening to challenge the point and add an important fact that might have been omitted. Try to steer the other panelists to comment on what *you* have said. If only weak answers are given to a question, look for an opportunity to add something that has some impact. If you are patient, you can find an opening in just about any panel's answer to make a strong point and again establish yourself as an expert.

Political debates are a good training ground for panel discussions. Politicians are highly polished speakers. They always try to have secure the strongest position on each issue that is discussed—after all, that is how they "win" the debate.

Establishing Yourself as an Expert

Almost all discussion panels comprise a collection of differing viewpoints and opinions. Your real goal should be to establish yourself as an expert.

By establishing yourself as an expert, it naturally follows that your company, technology, product, and perhaps other things that you represent will be perceived positively. If the audience does not see your product or technology as "best," at least now they probably feel it is worth investigating. And indeed that is the result that you are seeking.

Do not feel that you are not qualified to establish yourself as an expert. By preparing well and fielding questions with properly timed responses, the audience will perceive you as an expert, and that perception is what you are looking for. Besides, you are *much* more of an expert than those in the audience because most of them probably did not prepare at all.

Make Your Side Look Good

You can do things during answering questions to make your side look good. As mentioned earlier, quickly fielding questions that you can answer well, and playing off other responses will make you look good and also establish your position as an expert.

Steer questions your way, either from the audience or from other panelists. You can answer a question by touting a benefit of your position that will require panelists from other positions to defend themselves. For example, if your product has a particular feature that others don't, you can stress applications that would benefit from the feature. This will require those whose products lack that feature to defend its absence. You can then continue from their answer by listing additional benefits, which will leave them in an unfavorable situation.

However, you should not degrade others' positions or products, nor be explicit about their negatives. Always play off of your positives; stress pluses. You can even do this about features that other products have. If you bring up the feature first, it will appear to the audience that others simply copied your good idea, regardless of who did it first.

A number of years ago, Shell Oil Company ran a long series of commercials stressing the benefit of their gasoline additive Platformate™. In the commercials, a car with a large glass bottle of gasoline (without Platformate) mounted on the rear would drive until it ran out of gas. At this point in the road, a paper barrier would be erected. Then, the car would get its gasoline bottle filled with Shell gasoline with Platformate. The car would, of course, pass through the barrier clearly showing the benefit of Platformate.

What the viewers did not know is that virtually all gasolines had nearly the same additive as Platformate and performed better than cars burning gasoline without the additive. Even so, the audience perceived Shell as giving extra mileage because Shell was the first to tell us about it and in a very explicit way. Be positive about your product, without being negative about the competition, and you will win the audience's confidence.

DESIGN REVIEWS

Most technical people dread design reviews. However, with a little thought and preparation, a design review should not be much more difficult than a technical discussion with a friend. If you bear in mind that you know more about the design than anyone else in the meeting, you can relax and just present the facts. Typically, a design review consists of one or more lead engineers presenting a technical design progress report to superiors within the company, or prime contractors or customers outside of the company.

Design Goal

In preparing for a design review, set a design goal for your preparation—just as you have done in the other chapters. For a design review, the design goals for your preparation are two-fold: First, your primary purpose is to bring the attendees up-to-date on the progress of your design; second, you will want to establish your credibility, which will make the primary purpose much easier.

Your preparation will be similar to that for the panel discussion, but not as extensive. Since you are very close to the design, you will already have most of the knowledge. You must organize your knowledge, however, to be able to present your data and answer questions effectively.

Start preparing by trying to determine what information you will need to present and what questions might be asked. Try to answer as many of these questions with your presentation as you can. This way, you will not only establish your credibility, but you will decrease the number of questions that you will have to field later (and these questions are more difficult than making the presentation).

Review your design and your notebook to remind you why various decisions were made. You can just about bet that the participants will want to know why you made major design decisions. Remind yourself, in detail, why other solutions were rejected. If you have had major problems during the design, review the problem and how you arrived at the solution. Again, remind yourself of what the other solutions were that you considered and why they were rejected.

You will probably also want to ask other engineers and technicians who have been working on the design for input. As with the panel discussion, their viewpoint can often provide you with additional insight and other topics that should be presented.

Once you have done your preparation, gather your information together in a form that fits your particular design review format. If you will just be speaking, notes on $8\frac{1}{2} \times 11$ paper or notecards should work well. If your presentation is more formal, viewgraphs or 35 mm slides should be prepared. Chapter 9 details making and using visual collateral.

If you do use visuals, use them to your advantage, not just to have something to project on the screen. Showing a schematic is generally not too useful. The other attendees probably have not seen a schematic in a while, and are not interested in that level of detail anyway. Block diagrams, flowcharts, and other tabular material is much more effective and will help the others understand the design more easily.

Whichever presentation method you use, you will offer to answer questions following your verbal presentation. You should be able to field most every question if you have prepared well. When you answer, answer directly and honestly. If you stammer or have to pause to remember some fact because you did not review your own design well enough, you will hurt your credibility.

By all means, be honest with your answers. Do not try to cover up a problem you might have had, and if you do not know an answer, just say "I do not know." No one knows all the answers, and you will be more credible if you are honest.

For example, if you are asked if your disk controller would have been simpler if you had not used error correction, you might answer "Well, I don't know. We really didn't consider that because we felt that error correction was essential." Or, if you are asked if a larger bolt would be better, you could say "I don't know, we didn't try a larger one. However, our analysis showed that this size would be adequate, and should cost less than a larger one."

BOARD MEETINGS

Board meetings are also similar to panel discussions and design reviews, except that the level of technical content will be less. This is why these meetings are often the most dreaded of all: You must leave your own technical turf and enter the world of the board members, which is usually finance, marketing, and management.

Design Goals

The design goals for your board meeting are exactly the same as for a design review: To bring the participants up-to-date on the design, or department, and to establish your own credibility. Your preparation will be somewhat different because the participants will be different.

Start preparing with the technical side. You will feel most comfortable here, and it will help you by giving you a good foundation from which to build. Your technical presentation should be at a very high level and have few details. Block diagrams, flowcharts, and tables are okay, but they must be very general. The block diagram of a radio receiver in Figure 1 is fine for a design review, but for a board meeting, it should be reduced to that shown in Figure 2. Try to use analogies in describing the technical content so that the nontechnical members will understand and follow.

I have known people who went into a board meeting with the thought "I'm going to show Mr. Jones just how little he knows about how our product works." While this may give you some self-satisfaction, it probably will not promote your credibility nor will it ingratiate you with the board. If you strive to ensure that even the most nontechnical member has a basic

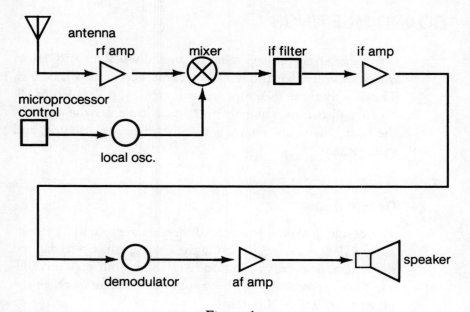

Figure 1
This block diagram of a radio receiver is too detailed for a board meeting.

understanding, you'll be much more successful—and it will not hurt your future any either!

To prepare for any nontechnical content, start by finding out about the board members and their background. You can get this information from your company's annual report or by asking a manager or vice president within your department. By researching their background, you can more ably anticipate questions. If a member has a particular specialty or market experience, include something directed at him or her, and be prepared for some questions that may be asked.

For example, if one of the board members has a financial background, try to provide some financial data, or at least be

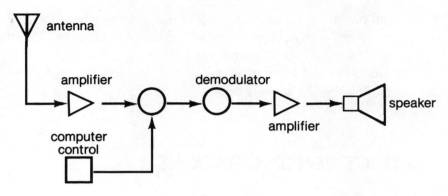

Figure 2

This less-detailed block diagram more appropriately suits the audience at a board meeting.

prepared to answer financial questions. If you are designing a computer printer and one of the board members has experience in medical electronics, you might point out the medical applications, or at least prepare yourself for the types of questions that might be asked. Think of how you would behave in a stockholder's meeting of a company that you own stock in. Although you would be interested in their profit and loss, their competition and so on, you would probably have a keener interest in their research and development spending, what new technologies they are investigating and how they compare to competitors in terms of technical leadership.

The participants in your meeting are analogous to this. You would be disappointed at the stockholder's meeting if no one could answer your questions; the same applies to your presentation.

It's important to be able to field the nontechnical questions. Many of these questions will be business related. After all, the board's main function is *not* how the technical department runs, but how the company runs its business. So, you must anticipate questions such as "How much will this cost to develop," "Why will it take so long to design?", and "How soon can we get it into production?" As always, answer directly and honestly, and if you don't know the answer, say so.

PRODUCT DEMONSTRATIONS

Of all types of technical speeches, product demonstrations are the most fun for me—unless the product doesn't work. If you have prepared well, this should not happen. I like product demonstrations best because it is a chance to show off your work. Everyone likes "warm fuzzies," and you should get plenty of them during a demonstration.

Design Goal

For a product demonstration, your goal is to show as much applicable functionality of the product as possible *within the constraints of the demonstration*. Product demonstrations are usually given either to your company's salesforce, or directly to a customer. This means that when you are finished, the salesman or customer should have enough understanding not only to use the product, but to understand why it fulfills his need as well.

"Constraints of the demo" is an important phrase. I have had many products demonstrated to me that were not very effective because little effort had been made to show me the functionality of the product. Few products can be effectively demonstrated without some test equipment. If it must be highly portable, it is worth the effort to build a specialized piece to aid the demonstration.

I once had a digital, storage oscilloscope demonstrated to me. I had told the sales engineer beforehand that I was especially interested in its ability to store data and then to perform mathematical operations on it later. In his demonstration, he was not familiar enough with the product to demonstrate much more than the basic functions and store a small amount of data. I did not buy the 'scope, but chances are that I would have if the sales engineer could have made it operate.

To demonstrate a faster disk controller board, one company used a computer and displayed several marvelous graphics screens. This was very colorful, but without a comparison, how could anyone tell that it was faster than another controller? Later the company changed the demonstration to be more effective. Instead of one computer, two were used: One with the company's fast product and one with a competitive product. Instead of displaying pretty pictures on the screen, both systems displayed a graph of the number of disk accesses or the data retrieved since the demonstration began. This was much more effective. You must include whatever test equipment is necessary to effectively demonstrate your product.

The first preparation you should do before a demonstration is to check out all of your equipment and be sure it works. You should do this before you leave the factory, and again if possible before you visit the customer or talk to your salesforce. If you have traveled to give the demonstration, try to exercise the equipment in your hotel room prior to making a "live" demonstration.

Demonstration to the Salesforce

New equipment is generally demonstrated to the salesforce so that they can effectively sell it. You should begin your demo with some design background. Why was it done the way you did? What did you do different from your competition and why is it better? How is it easier to use?

Next, extract the basic features and demonstrate them in detail. Most of the salesforce will be able to remember only a handful of details about a product, so be sure you demonstrate the most important ones first. This is not a criticism of salespeople, but rather an observation. They are not usually too technical, and they usually have several different products to sell so they must concentrate on only a few, but very pertinent, features.

Make sure they can "turn the product on" or "get it going." As trivial as this sounds, imagine how well the rest of the salesman's pitch will go if he cannot get the product going to begin with. Give them enough detail to get the product going and have it *complete* some task. This will give them a tool to do a basic demo to a customer.

After you have presented the basics, then you can go into more detail about those features and cover the more advanced features. Some of the salespeople will not get too much from this, but others will so do not get discouraged.

Demonstration to a Customer

The very first step in preparing for a customer demonstration is to talk with the salesperson handling the customer. Your function in giving the demonstration is to *augment* the salesperson by being the technical expert. You will want to learn what the customer's application is for your product. Learning this gives you the opportunity to suggest specific ways to use the product in his application.

Find out what the salesperson believes are the major selling points of your product. You should tailor your demo to highlight these features for the customer. Additionally, you may be able to suggest other important features to your salesperson to help him or her make the sale.

When you actually give the demonstration to the customer, follow the same procedure as you did with the salesforce: Give a background, a basic demo, and then move on to the more advanced functions. Your basic demo in this instance should be matched to the needs of the customer so that you will show him up front how your product can solve his problem.

The most important thing for you to do in a demonstration is to let the customer interact with the equipment. Demonstrate a feature, and then let the customer try it. Let him know you trust him, and let him get comfortable with the product. The more comfortable he feels, the more apt he is to buy the product. You'll know the customer is interested if he begins suggesting features to demonstrate. If he does, talk him through and let him do the actual demonstration.

Finally, answer the customer's questions directly and honestly. If you do not know an answer, tell him so, but promise to get the answer and get back to him. Give him a commitment date that you will get back to him with the answer, and stick to it. This is a touchy point because failing to get back to him when you promised can invalidate the whole product demonstration.

SALES TALKS

Many technical people are asked to make a call with a salesperson to help with the technical details. Application engineers' main job is exactly this. A product demonstration is one type of this function, but more often, your function will be to answer questions, provide technical details, and support the salesman. This, really, is the design goal for a sales talk: To provide technical *back-up* for the salesperson in a sales environment.

I emphasize back-up because you are the secondary person in this role. In order to prepare properly, you must meet with the salesperson prior to your meeting with the customer. Determine beforehand what points you will contribute to, and what points to stay quiet about. You should also arrange to use some sort of signal, such as a wink or smile, when the salesperson feels you should wrap up what you are saying so that he or she can assume their primary role once again.

Follow the salesperson, like a quarterback. The salesperson is the customer expert—that's their job. They have met with the customer before and have determined what the customer needs, how your product can fulfill the need, and what the best selling approach is. They have also determined when your help is needed.

There are several problems that arise with a technical person in the field. I am going to be very blunt, and I have been there too, so don't feel offended.

1. The technical person talks too much or volunteers too much information.
2. The technical person goes beyond the technical capabilities of the customer.
3. The technical person "unsells" the product.

It's very easy for a technical person to talk too much during a sales talk. You are likely to get along quite well with the customer, and talk on and on. This is detrimental to the sale, though. The solution to this problem is in the meeting with the salesperson prior to meeting with the customer. You should ask exactly how much to say about each topic that the salesperson wants your help with. Set up and use visual signals so that you can get things back in the hands of the salesperson. And, don't volunteer information about every topic. The salesperson has a strategy, and your intervention may disrupt it. If he wants information from you, he will ask. You can also discuss this beforehand, and offer to comment whenever he or she would like you to do so.

Whenever you talk to a customer, you must determine his technical level of expertise. I favor the direct approach: "How technical are you?" often will bring an honest answer from which you can gauge your level of presentation. Watch for visible signals that the customer either does or does not understand what you are saying. You can even ask if he understands a given concept. Your salesperson should also be able to give you an idea regarding the technical competence.

The final problem is unselling. The most common occurrence is when the technical expert responds to a customer along the lines of "Oh yeah, we know that's a problem and our next generation product will have this . . ." It's going to be next to impossible to sell your current product to that customer at this point, plus you have made a promise to the customer of delivering this new product. Sometimes, a direct promise is given such as "we can customize this in three weeks." Technically, this is probably correct, but when can you work it into your resources' time to get the three weeks? It may be six months from now. If you promised three weeks, the customer (and the salesperson) will want it three weeks from when he ordered it, not three weeks from when you get to it.

Your salesperson's job is to sell. Don't cancel his work or accidentally defeat it. Remember that the salesperson's pay is based on sales. If you kill a sale, you are not likely to be too popular with the salesforce.

SUMMARY

Most technical people have to speak technically from time to time. For most of us, there are a million things that we would rather be doing, but the benefits can be very rewarding. You can improve your stature as well as that of your company.

The key to giving effective technical speeches is to prepare well for them. This entails research, questioning, and being open and honest. The different types of speaking detailed in this chapter—technical speeches, panel discussions, design reviews, board meetings, product demonstrations, and sales talks—are all related, but you need to prepare for each in a different way.

As with writing, work from a specification, set a goal, and work toward meeting that goal as well as you can.

CHAPTER
9

DELIVERING THE SPEECH

Since as much as 90% or more of getting your message to the audience depends on the way that you deliver your speech, it's critical that you put as much effort into preparing for the delivery as you can.

*A*lthough it may not seem so at first investigation, a significantly larger portion of the message is transmitted in the delivery of a speech, rather than its content. Recent research has shown this dramatically. For example, one such study showed that only 7 percent of the message was conveyed with the actual words, 38 percent from the speaker's tone and vocal inflections, and *55 percent* from the visual content. The visual content includes not only transparencies, slides, and other graphics, but facial expression, gestures, and movements as well.

Since as much as 90 percent or more of getting your message to the audience depends on the way that you deliver your speech, it is critical that you put as much effort into preparing for the delivery as you can.

More than a third of your message transmittal depends on not what you say, but *how* you say it. You've listened to a speaker who spoke with a meek, wavering tone. Immediately, you recognized that this was probably the first time the speaker had given a speech, and that the speaker was scared stiff! You may have also noticed that you gave less credence to what that person had to say than someone else who spoke with authority. The confidence of the speaker is somehow reflected by the listener.

Therefore, when you speak, your tone should exude confidence. The audience needs to be convinced that *you* are confident in what you are saying before they can be convinced that what you say is correct. Just because you are confident doesn't necessarily mean that you will convince the audience; however, if you do not sound confident, it will be next to impossible to convince anyone else. This is another good reason not to read your speech verbatim. Actors do not read lines, but memorize them so that they can concentrate on the delivery instead of the content.

SOUND WITH IMPACT

It is also important to sound excited about your topic. Sometimes, the excitement is reflected more as a keen interest. Regardless, if you do not sound interested or excited, you will never get your audience involved. To do this requires practice and familiarity with your topic.

A perfect example of this effect is a sportscaster. You have certainly noticed that the local sportscaster is generally not nearly as interesting as the major network's commentators. National sportscasters have gotten their positions because they have a talent that made them better than their competition. That talent is the ability to involve and excite the audience. Part of why they can do this is that they spend a lot of time preparing. They do not have to think "number 80 caught that pass, let's see, that's Jones"; they know the team well enough that they recognize without thinking that Jones caught the pass.

They also are interested in, and enjoy what they are doing. It is clear that they are genuinely impressed when an athlete makes an extraordinary play and are disappointed when one fails. Sports play-by-play commentators are among the best professional speakers in the world. The reason why is that they are confident, interested, and excited about the topic they are speaking about.

THE VISUAL SIDE OF SPEAKING

If you want to be an effective speaker, you must have good visual communication. Even if you have fantastic information, and even if you are excited and vibrant about your topic, if your visuals are not effective, your speech will not be very effective. Discoveries made about sensory perception in humans proves that we depend most heavily on our sense of sight. You must appeal to this sense in such a way that your message will be remembered.

The most effective way to communicate your message is to couple visual displays with your speech. You should make every effort to use visuals anytime you have to speak in public. The best visuals are overhead transparencies, or foils, and 35 mm slides. Other types of visual aids are sometimes dictated by circumstances, but are seldom as effective.

In addition to augmenting your words with visual communications, you must incorporate gestures, facial expressions and eye contact, and movements. This is especially true in presentations in which other visual aids are not feasible, such as during a panel discussion. Continually scan the audience as you speak and make as much eye contact as possible. Exaggerate facial expressions from time-to-time so that even the back row of the audience will be able to see them. Gestures with your hands are also important. Simply waving your hand in front of you can cement an otherwise weak point. Moving around the stage, in situations that allow movement, keeps your audience's attention focused on you rather than wandering. One important note however: Stay out of your audience's way so they can see your visuals.

Tom Peters, co-author of *In Search Of Excellence* is a fantastic person to watch speaking. He incorporates all the elements of positive presentation mentioned here. Peters is clearly interested in what he talks about and is unquestionably confident! Beyond that, he gets *excited!* Peters yells and screams and laughs. He waves his arms and paces the stage. He uses large, simple, and readable 35 mm slide projections. And, as a result, it's virtually impossible not to get excited with him, and come away saying "he's right!" *Tom Peters communicates 100 percent.*

DESIGN OF YOUR DELIVERY

Your design goal with a speech of any kind is to communicate effectively to your audience. To accomplish your goal, you must do three basic things: (1) Win their confidence, (2) get the message across, and (3) get the audience excited. Chapter 8 dealt with winning your own confidence and winning the audience's confidence. The balance of this chapter focuses on getting the message across and getting the audience excited so that the message is retained.

Practice

Practicing your speaking is the most important step to becoming a good speaker. The best way to practice is to speak often. If you really want to become a dynamic, well-honed speaker, you must practice in actual speaking engagements, in front of live audiences. You can join a speaking organization, such as Toastmasters. You can (and should) make presentations within your own company whenever possible. Most companies would love to have someone from each department tell other departments about what they do.

Many companies encourage "brown-bag lunches" where someone puts on a small program, during lunch. The topic is just about anything; a hobby, a craft, an interest, a sport, or even a work-related topic. Schools and organizations such as scouting beg for people to make presentations to children. You can find ample opportunity to serve your apprenticeship.

You also need to practice your particular speech to be able to present it effectively and confidently. Practice is boring and takes up time that you would probably rather be spending doing something else. The truth is, most people do not practice their speeches much. However, some practice is imperative, and the more practice you get, the better your presentation will be.

I once traveled around the country with a new product making presentations to major customers. In all, I made over a dozen presentations of the same material. While making these, the practice led to omitting areas that were not important; adding areas that frequently came up as questions; trimming the fat; and finally, the presentation was even reduced from two days to one and one half. The practice had yielded a more effectively communicated message in a more time-economical format.

If you do nothing else, recite your speech aloud. Doing this while driving to work is effective use of your time and is not as potentially embarrassing (provided you don't participate in a car pool!) as trying to recite it in a crowded area, such as an office. You should make an attempt to time your speech after you have given it to the freeway a couple of times to be sure that your estimate was accurate. If necessary, you will still have the opportunity to add or delete material as time requires.

You should try to taperecord yourself reciting your speech. You may be able to do this during your drive into work, and not require any more time. When you listen to your speech, listen more for inflection and presentation than content. By this point, you should have your content down pretty well, but it will be the first time you will have heard what the audience will hear. When you come to points where you need emphasis or special inflection, mark it on your index card. Often, a simple underline will be enough to remind you to emphasize a point.

Many companies and schools are also using videotape to enable the speaker to improve his delivery technique. Since so much of your message is visually communicated, this technique is most likely the best practice you can get. If you have access to a videocamera, you can even videotape yourself at home. As with a taperecording, don't concentrate on the information, but on the delivery, especially the visual portion. You can make notes on your cards to indicate points where gestures, facial expressions, or movements will aid in making your point.

You may also want to arrange to present your speech to all or part of your company. Often, others within your own department or related departments would be interested in learning what you have to say. While the atmosphere is usually more friendly and easier to speak within, coworkers and friends are usually more likely to give you constructive feedback than strangers. Be open to their comments and your speech will benefit.

If your speech incorporates visual aids, step through each one. Think, or better tape, what you will say about each. If you use overheads with cardboard frames, you can make notes on the frame. Check especially that you have them in the correct order, and that the order you have them in is the best order. I have often rearranged the order of my speech upon closer thought about the visuals. I have also saved myself from embarrassment by finding out-of-order visuals *before* I got in front of an audience.

No matter how much, or how little practice you can fit in, do some. I have seen many speeches that had good material and were presented by an intelligent and interesting person, but were ineffective because the speaker was not familiar with his own material. If for no other reason, practice to save yourself embarrassment!

Organize

The more organized you are, the more confidence you'll have in your presentation, hence, the better your presentation will be. This is even more important for panel discussions or board meetings when you need to locate information quickly.

Your first step is to begin getting organized is to arrange your notes. If you are making a technical speech, you probably have already done this, so a simple check to confirm their order and that none is missing should do. If you are preparing for a panel-type presentation, you should index your notecards in some way. An easy way to do this is to use small stick-on notes, such as the Post-It™ series by 3M. Print a key word on each, cut as small as you can, and stick it onto the top card of those being indexed under that heading. Then during your presentation, if you need to refer to your notes for information, you can locate specific data quickly enough to be useful.

Next, organize your visuals. If you are using 35 mm slides, number them in the final order that you have decided upon. It is also a good idea to put your name and address on the cardboard mount in case you lose any of them. The best way to transport your slides is in a Kodak Carousel. Virtually every facility providing a slide projector uses a Carousel-compatible unit, but by all means check with them before presuming this is what you will have available. By transporting them in a Carousel, you should never have to insert them into a Carousel and run the risk of disrupting the order or inserting the slides backwards. You can simply drop your own Carousel right into the projector and begin. Another help in organizing 35 mm slides is to produce a company logo slide. Use this as your first slide to help you find the beginning of your presentation and as a focusing aid.

If you are using foils and an overhead projector, your organization will be aided if you mount your foils in cardboard frames. You can then use the frames in the same way as the 35 mm slide mounts, to put your name on and to number them so that you can quickly reassemble them into the proper order should they become shuffled. With foils too, you can have a logo made to use as a focusing aid.

As your final preparatory step, set up a timetable for the day of your presentation. For most speaking engagements, the organizers request you to arrive anywhere from 30 minutes to an hour ahead of schedule. This is a good idea because it gives you time to unpack your materials, check in with the management and relax, which will take you mind off your speech as well. Most places have a speakers' lounge with a slide projector or overhead projector in case you want to review your materials one more time.

Determine a timetable for the rest of your day as well. If you aren't staying in the hotel that the conference is in, be sure to allow ample travel time, including a buffer for traffic.

VISUAL SPECIFICS

It would be difficult to overemphasize the importance of your visual presentation, bearing in mind that it's responsible for transmitting over half of your message. You can, and should use visual aids in just about every type of presentation except a panel discussion. Regardless of the type of visual you use, keep each one simple and limit them to one topic per page. For projected images, try to stay at 5 or 6 lines of text or fewer. Remember that you will want the back row of the room to read your image easily.

Some presentations have limits on what type you can use. There are six basic types of visuals that you can use.

35 mm Slides (35 mm) have been discussed in some detail already. For most presentations, 35 mm slides are the most effective form of visual communication if the conference is set up to allow their use. 35 mm slides are professional looking, effectively use color, and are compact to transport. Cost is moderate since their production will generally require you to hire the assistance of a graphic artist. You will also need to allow enough

time for the slides to get produced. You should not begin working on slides any less than 4 weeks before your speaking date, and two months is preferred.

Transparencies or foils are also quite effective and commonly used. Transparencies have the advantage of being producible with an ordinary xerographic copy machine. However, for a speech at a show or conference, professionally produced foils are much more effective. You can approximate professional quality with a desktop publishing system and a laser printer, but color is still difficult.

You can add color to black and white graphs, charts, and tables by using transparent colored mylar tape. You can buy the colored mylar in either sheets or tape. The sheets are much easier to use because you can cut any size or shape with a utility knife and then transfer the mylar directly to your foil. Adding spot color to your foils can make a big difference even with departmental presentations. The cost of a package of colored sheets is about $15.

Another effective way to use foils is to make a multiple-overlay foil (Figure 1). Different foils are mounted to different edges of the frame, then each is folded over the base foil one at a time as your presentation progresses. This works best if your foils are mounted in a frame, but don't let that limit you. You can use the multiple overlays to add either spot color for emphasis, or lines of text as you progress through your monologue. These are easy to do with a strip of masking tape, and are very effective in getting your message to your audience.

For example, if you have a pie chart showing three ways to break something down, you might have a basic black and white image depicting the pie. Then, from each of three sides you can fold over the overlay that highlights one of the three slices with a different color, and speak about each as you fold them in. Alternatively, you can have a basic foil with text and

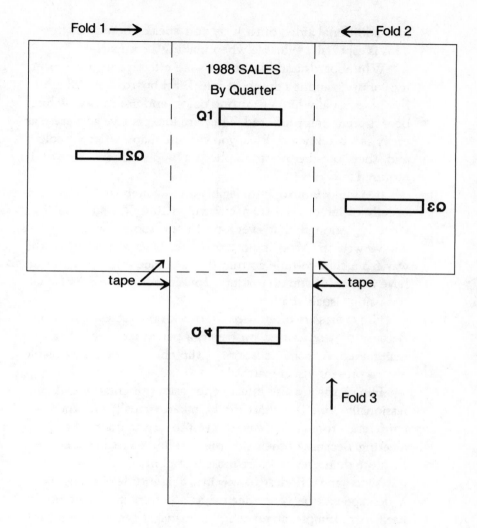

Figure 1

A multiple-overlay foil enables you to add material to the foil as you speak, or highlight sections of the foil as you cover that particular topic.

fold additional lines onto it as you speak about each topic. This is especially valuable when making a comparison.

Whiteboards and blackboards are effective in small meetings of up to about a dozen people. Blackboards are difficult to read, are usually limited to one color, and are messy. Whiteboards are much preferred. They are inexpensive and erase as easily as a blackboard. Plus, you can use many different colors and they can be photographed if a permanent record is required.

It is important to write legibly on a whiteboard. Printing is usually easier to read than cursive writing. You also need to write big enough that everyone in the room can read what you've written. Most importantly, you have to get out of the way from time-to-time so that the audience can see what you have written. If you do not let them see, you are worse off than not using visuals at all.

Flip charts are often used for quick notes at sales meetings. Because they're written on paper, a permanent record can be maintained. At some locations, a flip chart is the only feasible way to present visual material.

Flip charts are also effective for sales presentations. A professionally done flip chart can be taken virtually anywhere—a conference room, an office, or even to a restaurant for a lunch meeting. Because of their size, they are limited to an audience of no more than a few, and are most commonly used one-on-one.

Videotape is finding its way into a multitude of applications. Videotape can be very effective as a presentation medium by itself. For example, an engineering manager can make a technical presentation on tape that each salesman can get a copy of. The salesman can have an application engineer with him by just playing the tape. It is not as good as a live body, but it can be effective.

You can also use videotape as an audio-visual aid. Rather than having the tape make the entire presentation, you can use vignettes from a tape to stress important points or make illustrations that are impossible to do live.

Computer screens are being used more and more as visual aids for verbal presentations. You can use a computer directly when your audience is limited to a very small number. But devices are also available that allow you to project the image from the screen onto a movie screen so that a large audience can see what is actually taking place on the computer. Other devices can allow you to copy the screen onto paper for distribution.

USING VISUALS

Now that you are aware of the different types of visuals that are available and when you might use each of the types, let's look at how you can most effectively use each of the types.

35 mm Slides

For most presentations, 35 mm slides are the most effective visual aid that you can use, for the reasons mentioned earlier. Getting 35 mm slides produced is difficult from the perspective that you will most likely have to work with someone outside your own company. This requires that you allow enough lead time to get the slides produced. This has a benefit in that it forces you to get your presentation finished well in advance, and prevents you from waiting until the last minute.

When you have your slides made, have the artist insert a small copy of your company logo in the bottom right-hand corner of each slide. This serves two purposes. First, it will add some exposure for your company each time the slide is projected. This can be especially valuable during a technical presentation when you are speaking about technology rather than your product. Secondly, standardizing on the bottom right-hand corner will enable you to insert quickly your slides into the Carousel without getting them upside down or backwards. If you have to transfer your slides into a Carousel, this is especially helpful.

When putting together the material for your slides, be sure to keep it fairly simple—do not get overly detailed. Most slide projectors, even those with zoom lenses, have a short focal distance. In other words, they can only be moved back so far, so the maximum image size will be rather limited unless the projector has a special long-focus lens. You will want everyone in the room you are presenting in to be able to read the slide, so if you get too detailed, the printing will be too small to see from a distance.

Recall also that the visual is an *aid* to your presentation. Its purpose is to reinforce what you are saying and to provide a visual "hook" to help the audience remember what you have said. It should not be used to state everything you want to say so that you don't have to talk. There are two rules of thumb:

1. Cover only one topic per slide.
2. Use no more than 6 lines per slide if possible. Adhering to these rules will give your slides more impact, and make them a more valuable tool in getting your message across to the audience.

These same rules of thumb are adhered to for large, outdoor billboards as well. Next time you are driving past a billboard, notice how few lines of text are on the sign, but yet how effectively good ones communicate the message. One of the better ad campaigns I have seen was a series of three billboards for the opening of a new Carson Pirie Scott and Company department store in the midwest. The first sign, erected about three months prior to the opening said, simply "On Your

Mark." About two months before the opening the signs were replaced with ones reading "Get Set." By this time, they had everyone's attention to the point that people were even having conversations about what the signs meant. Finally, about two weeks before the grand opening, the signs were changed to "Carson's Grand Opening, June 10." The audience was hooked. Simple can be much more effective than detailed.

Another effective technique with slides is to "add-in" material with successive slides. For example, your slide might read:

Benefits of Winchester disk drives:

1. Increased storage capacity

2. Faster data access

3. Greater reliability

You could simply project one slide with this list. Perhaps a more effective way of presenting this material would be with *three* slides. The three, in sequence would read:

Slide #1

Benefits of Winchester disk drives:

1. Increased storage capacity

Slide #2

Benefits of Winchester disk drives:

1. Increased storage capacity

2. Faster data access

Slide #3

Benefits of Winchester disk drives:

1. Increased storage capacity

2. Faster data access

3. Greater reliability

Highlighting the benefit in color as you add it also increases the effectiveness.

Foils

If you cannot use 35 mm slides, or if your presentation is more informal, you can use foils (transparencies) with an overhead projector. There are two ways to make transparencies: do them with a xerographic copier (photocopier or laser printer), or have them done professionally. Most of the transparencies that you use will be done yourself, or at least inhouse, using a photocopier. These are suitable for most informal presentations. If you are giving a formal presentation, it is worth the extra time and money to have them done professionally, and in color.

If you do the foils yourself, try to use a typewriter or printer with extra large printing, or have the drafting department letter them with templates or pressure-transfer letters onto plain white paper. Having your information typeset will look even better. You can then use transparency material in a photocopier to make the foils. If you can, mount the foils in cardboard frames to make handling easier. Framed foils are also easier to store and reuse later. Using colored tape on the foils can add the accents that will help your audience remember the key points.

The same considerations apply to foils that applied to 35 mm slides. You need to letter large enough so that the whole audience will be able to read the information; stick to only one topic per page; and try to use six lines or less per foil.

Whiteboards and Flipcharts

Whiteboards and flipcharts are used in similar situations. They are used for informal presentations when only a few people are present, usually a dozen or fewer. The reason for this is because it is difficult for everyone to see if many more than a dozen are present.

Whiteboards are handy in that they may be erased and reused many times during the presentation. You can also use different colors of ink for emphasis. Remember the basic rules when using a whiteboard though; print large and legibly, and do not get too detailed. If you change topics on one whiteboard, try to separate the topics with a line so that you can shift your audience's attention.

Flipcharts can be used when you do not have access to a whiteboard, such as in a hotel, or when you want to save what is written on the charts. You can also prepare the flipcharts in advance so that they will be neater. Often, flipcharts are used to make action-item lists that are constructed at a meeting. With both flipcharts and whiteboards, the most important factor is to stay out of the way so the audience can read what you have written.

Videotapes

By all means, have a videotape done professionally. Unless your company has a good dual-drive editing machine and an experienced editor, trying to do the tape yourself will result in a poor visual aid. Making videotapes is not an inexpensive proposition; however, they can be very effective and can enable the presentation to be shown to many audiences that would be otherwise impractical.

Videotapes that only show the speaker making his presentation are ineffective. In fact, you should keep the speaker off camera except for cameo shots. The speaker's face isn't the main message of the tape. You have access to a very powerful medium—use it to its fullest capability. Include action shots when possible, show actual demonstrations of equipment being used in typical applications, and show a lot of people. People give the viewer someone to identify with.

Finally, keep the length as short as you can and still make a clear point. You will start to loose the attention of the viewer after only a few minutes. A ten to 15-minute video is adequate for most topics. Be sure to use a lot of color.

Using a Computer

Several devices introduced in the past few years have made computers great tools for making visual aids for presentations. The desktop publishing systems allow you to create many fonts and sizes. When coupled with a laser printer, you can make foils directly from the computer. The laser printer uses the same xerographic technique as a photocopier, so you can bypass the printing stage entirely.

You can also make 35 mm slides directly with a computer. There are devices that interface with the computer which will make 35 mm slides from the screen. These are fairly expensive—however, there is an alternative. You can make the templates for the slides using a desktop publishing system. Then record these on floppy disk and take the disk to a service bureau that will make the 35 mm slides from the disk. The cost of this service is usually $5 to $15 per slide.

If your computer is an integral part of your presentation, you can do one of two things. A projection device will allow you to project the image from the screen directly onto a movie screen. This is extremely useful when training people on a particular software package. Alternately, there are CRT reproduction boxes available that will use photocopying techniques to print the screen on a sheet of paper exactly as it appears on the screen.

All in all, you can use a computer with a desktop publishing system to make just about any kind of visual aid that you might need for a verbal presentation.

MAKING THE PRESENTATION

Making a good presentation requires coordinating the research, writing, and content of the speech with the visual aids and your vocal inflection and tone.

For a technical speech, remember that you are the expert on the subject that you are covering. Research, then practice enough that you have a feeling of confidence. Speak with intensity and interest. Use the most effective visuals that the situation and your budget allow. Do not use too many visuals, and by all means keep them simple. Use the visuals to drive home the most important points and to clarify those that are difficult to present orally.

A panel discussion requires you to effectively research and then index your research so that you can access it quickly. Since you are almost never able to use visuals, you must use facial expressions and gestures, coupled with strong inflection and intonation to augment the oral content when you speak.

Design reviews and board meetings are similar to panel discussions. However, you can use some kind of visual aid in these two types of speeches. By all means, use whatever you can. Visual aids will not only help you drive home your point, but will make you look more professional as well.

In a product demonstration or a sales presentation, you will already have a visual aid—the product. Do not try to upstage the product because that is what the customer is interested in. If you have some graphs, charts, or specifications, incorporate them. Small, professionally prepared flipcharts, if they are not too lengthy, are very valuable too, as are videotapes.

The way you present your material and yourself can be more important than the actual content of what you present. More than a third of getting the message across depends on the way you speak; confidence and interest will score highly. Over half of driving home your point depends on the visual help you provide to the audience to keep their attention and aid in their understanding. Even with the best material, your speech will not be effective if you do not present it well.

INDEX